T0332278

THE SYNOPTIC PROBLEM AND STATISTICS

THE SYNOPTIC PROBLEM AND STATISTICS

Andris Abakuks

CRC Press is an imprint of the
Taylor & Francis Group, an **informa** business
A CHAPMAN & HALL BOOK

CRC Press
Taylor & Francis Group
6000 Broken Sound Parkway NW, Suite 300
Boca Raton, FL 33487-2742

© 2015 by Taylor & Francis Group, LLC
CRC Press is an imprint of Taylor & Francis Group, an Informa business

No claim to original U.S. Government works

Printed on acid-free paper
Version Date: 20140903

International Standard Book Number-13: 978-1-4665-7201-0 (Hardback)

This book contains information obtained from authentic and highly regarded sources. Reasonable efforts have been made to publish reliable data and information, but the author and publisher cannot assume responsibility for the validity of all materials or the consequences of their use. The authors and publishers have attempted to trace the copyright holders of all material reproduced in this publication and apologize to copyright holders if permission to publish in this form has not been obtained. If any copyright material has not been acknowledged please write and let us know so we may rectify in any future reprint.

Except as permitted under U.S. Copyright Law, no part of this book may be reprinted, reproduced, transmitted, or utilized in any form by any electronic, mechanical, or other means, now known or hereafter invented, including photocopying, microfilming, and recording, or in any information storage or retrieval system, without written permission from the publishers.

For permission to photocopy or use material electronically from this work, please access www.copyright.com (http://www.copyright.com/) or contact the Copyright Clearance Center, Inc. (CCC), 222 Rosewood Drive, Danvers, MA 01923, 978-750-8400. CCC is a not-for-profit organization that provides licenses and registration for a variety of users. For organizations that have been granted a photocopy license by the CCC, a separate system of payment has been arranged.

Trademark Notice: Product or corporate names may be trademarks or registered trademarks, and are used only for identification and explanation without intent to infringe.

Visit the Taylor & Francis Web site at
http://www.taylorandfrancis.com

and the CRC Press Web site at
http://www.crcpress.com

... πᾶς γραμματεὺς μαθητευθεὶς τῇ βασιλείᾳ τῶν οὐρανῶν ὅμοιός ἐστιν ἀνθρώπῳ οἰκοδεσπότῃ, ὅστις ἐκβάλλει ἐκ τοῦ θησαυροῦ αὐτοῦ καινὰ καὶ παλαιά.

... every scribe who has been trained for the kingdom of heaven is like the master of a household who brings out of his treasure what is new and what is old.

... ikkatrs rakstu mācītājs, mācīts Debesu valstībai, ir līdzīgs nama kungam, kas izdod no sava krājuma jaunas un vecas lietas.

(Mt 13:52)

Contents

Preface

This monograph is an attempt to lay the foundations for a new area of interdisciplinary research that uses statistical techniques to investigate the synoptic problem in New Testament studies, with potential application to the study of other examples of sets of similar documents whose relationships are to be explored. Published in a series of statistics monographs, it is aimed primarily at academic and professional statisticians, for whom an introductory account of the synoptic problem and of the relevant theories, literature and research materials is provided. For those with no special interest in biblical studies or textual analysis, the core statistical material on the use of hidden Markov models to analyse binary time series may be of interest in its own right. The binary time series data sets used in this book, together with the relevant R code, may be found at `www.ems.bbk.ac.uk/faculty/abakuks/synoptic`. It is hoped that the book may also appeal to biblical scholars with an interest in the synoptic problem or more generally in the use of statistical methods for textual analysis, even if they have to omit the more technical mathematical and statistical aspects of the material.

I am grateful to my colleagues in the Department of Economics, Mathematics and Statistics at Birkbeck for their friendly tolerance of my research in biblical studies and especially to Ron Smith for his ready willingness to read through and comment on drafts of my work. Birkbeck granted me sabbatical study leave for the academic year 2013-14 to help me to complete the final draft of this monograph, for which too I am particularly grateful.

Over the years 1995-2003 I was a part-time student of theology at King's College London, where I was introduced to the academic discipline of biblical studies. During my sabbatical year, I have been stimulated by being able to participate in the research activities of the Department of Theology and Religious Studies at KCL. A number of former and current members of staff at KCL have been a source of inspiration, but Edward Adams deserves special thanks, first for suggesting to me the potential of the synoptic problem as an area for statistical investigation and then for his helpful comments on a draft of this work as it neared completion.

Through email correspondence I have received much encouragement from David Mealand and John Poirier, both biblical scholars who have also made significant statistical contributions to New Testament studies. I have been encouraged too by David Bartholomew, whose publications on the applications of probability theory and statistics to theology and biblical studies emboldened me to persevere in my own line of research. The editors and anonymous referees of my articles for the *Journal of the Royal Statistical Society, Series A*, have provided significant input and have helped to sharpen my ideas and improve my presentation of them. My thanks go also to Rob Calver and his colleagues at Taylor & Francis, who have made the task of preparing the manuscript go smoothly and efficiently.

But above all I would like to record my thanks to my wife Rūta, who encouraged me to study theology and has had to live with the consequences.

Andris Abakuks
Birkbeck
University of London

1

Introduction to the gospel texts and the synoptic problem

1.1 The *New Testament* gospels

Those who are familiar with the Christian bible will know that in the *New Testament* there are four gospels, the gospels according to Matthew, Mark, Luke and John. Each of them is in some sense a biography of Jesus Christ and tells essentially the same story about his life, death and resurrection, but each in its own way and from its own perspective.[1] When the texts of the first three gospels, those of Matthew, Mark and Luke, are looked at in parallel, it is especially noticeable how similar they are in the way that they recount particular episodes or sections of teaching, sometimes agreeing in their accounts almost word for word. These three gospels are known as the *synoptic gospels*.[2] The Gospel according to John is rather different in style and content, and the parallels with the synoptic gospels are generally not as close.

These four gospels, which came to be regarded as normative by the early church, are also known as the *canonical gospels*. There are other so-called gospels too, the non-canonical, apocryphal gospels, which have received considerable attention in recent years, but they do not form part of the *New Testament*. The most important of these is the Gospel of Thomas, which is a collection of the sayings of Jesus rather than in any sense a biography, and of which there is a complete surviving text in the Coptic language.[3] However, the non-canonical gospels do not fall within the scope of the present study.

There is no solid historical evidence as to who wrote the gospels, including the canonical ones, with which we are concerned here. The names traditionally

[1]There are a number of good introductions to the study of the gospels, for example, Hengel (2000), Stanton (2002), Burridge (2005) and Adams (2011). For the gospels viewed in the context of Graeco-Roman biography, see Burridge (2004).

[2]The meaning of the Greek behind the term *synoptic* is that these gospels can be "viewed together".

[3]Watson (2013) gives a wide-ranging discussion of the emergence and reception of the fourfold gospel canon and argues that the Gospel of Thomas is a surviving example from a genre of written sayings collections, precursors of which would have provided source materials for the writers of the canonical gospels. For the case that the Gospel of Thomas displays an acquaintance with the synoptic gospels see Gathercole (2012) and Goodacre (2012). For a general introduction to the non-canonical gospels see Foster (2008) and, for their texts, Ehrman and Pleše (2011).

attached to them, although going back to an early time in church history, are not regarded by most modern scholars as necessarily indicating their true authorship. Neither is it known where and when the gospels were written, although again there are some early traditions regarding this. For example, Irenaeus of Lyons in the second half of the second century, defending the fourfold gospel, writes (*Against Heresies*, 3.1.1):

> Thus Matthew published among the Hebrews a gospel written in their language, at the time when Peter and Paul were preaching at Rome and founding the church there. After their death Mark, the disciple and interpreter of Peter, himself delivered to us in writing what had been announced by Peter. Luke, the follower of Paul, put down in a book the Gospel preached by him. Later John the Lord's disciple, who reclined on his bosom, himself published the Gospel while staying at Ephesus in Asia.[4]

Papias, quoted by Eusebius in his *Ecclesiastical History*, III.39.15, is an even earlier source, who writes that "Mark became Peter's interpreter and wrote accurately all that he remembered ..." and that "Matthew collected the oracles in the Hebrew language, and each interpreted them as best he could".[5] These passages have often been cited and discussed as regards their precise meaning and implications.[6] Few scholars nowadays would accept them as historically accurate in all respects although recognising that they may contain some elements of truth.

The year AD 70 was a critical moment in Jewish and Christian history, when, after a Jewish revolt, Jerusalem fell to Roman forces and its magnificent temple, the focus of Jewish worship, was destroyed. Although there is no firm evidence, it tends to be generally accepted that the synoptic gospels were written at some time around AD 70 or within a few decades after that date. This was also the period when almost all of the last surviving eyewitnesses to the events described in the gospels, culminating in the crucifixion in around AD 33, would have been passing away.[7] After its fall and desolation, Jerusalem, where the first leaders of the Jesus movement had resided, could no longer remain as the main source of authority for the early church. Instead, the great cities around the shores of the Mediterranean, such as Rome, Ephesus, Antioch and Alexandria, emerged as the main centres of Christianity.

---

[4] Translation taken from Grant (1997).

[5] Although the dating is uncertain, Papias was probably writing early in the second century. The Greek text of the fourth century church historian Eusebius with a parallel English translation is available in Eusebius (1926). He also gives the above quotation from Irenaeus at V.8.2-4.

[6] For a recent discussion, see Watson (2013), pp. 121-131 and Chapter 9.

[7] For a scientific approach to the dating of the crucifixion, Humphreys (2011) may be consulted.

The Gospel according to Luke begins with the prologue:[8]

> Since many have undertaken to set down an orderly account of the events that have been fulfilled among us, just as they were handed on to us by those who from the beginning were eyewitnesses and servants of the word, I too decided, after investigating everything carefully from the very first, to write an orderly account ...

which in itself suggests, but no more than suggests, that this was not the first of the gospels to be written. However, considered in conjunction with the evidence of Papias and Irenaeus, Luke's prologue may be used in support of the view that his was the last of the synoptic gospels to appear.[9] Although it is not known for sure in what order the synoptic gospels came to be written, it does seem to be the case that the Gospel according to John, with its more developed reflection on the life and teaching of Jesus, was the last of the canonical gospels to be completed.[10]

1.2 Manuscripts, critical editions and synopses

The gospels, like the rest of the *New Testament*, were written in *Koine* Greek, the Hellenistic Greek that at the time was the commonly spoken language throughout the eastern half of the Roman empire, its *lingua franca*. The earliest surviving gospel fragments are found on papyri from the second and third centuries, but the earliest surviving manuscripts that contain a complete text of the gospels are the fourth-century *Codex Sinaiticus*,[11] now mainly in the British Library, and *Codex Vaticanus*, in the Vatican Library. Based on these and many other manuscripts, as described in Metzger and Ehrman (2005), New Testament scholars have produced critical Greek texts that attempt as far as possible to restore the original texts of the gospels and the other New Testament documents. Nowadays, the standard critical editions of the New Testament are the successive editions of the Nestle-Aland *Novum Testamentum Graece*, currently the 28th edition, Nestle-Aland (2012), commonly denoted by NA^{28}.

An important tool in studying the relationships between the gospels is the

[8] Here and later the translation of the Bible used for quotations from the gospels is the *New Revised Standard Version* (NRSV).

[9] For such a view, see Watson (2013), pp. 130-131.

[10] For a survey of views about the extent to which the Gospel of John shows knowledge of the synoptic gospels, see Smith (2001).

[11] A study of the earliest Christian manuscripts may be found in Hurtado (2006). A detailed description of the *Codex Sinaiticus* with historical background is given by Parker (2010). The Codex Sinaiticus Project website at `codexsinaiticus.org` is a superb resource for detailed study of the manuscript.

synopsis, where the texts of the gospels are laid out in parallel columns for easy comparison, either the three synoptic gospels or all four canonical gospels. Over the years various forms of synopsis have been produced, as reviewed by Greeven (1978). The ones most used in recent times are those of Huck (1949), for the synoptic gospels, and Aland (1996), for all four gospels. An English language synopsis of the synoptic gospels, that uses the NRSV translation, is provided by Throckmorton (1992), but the more recent English language synopsis produced by Crook (2012) translates the Greek text word by word, so as to provide a more precise comparison of the Greek of the four canonical gospels, and also exhibits parallels with the Gospel of Thomas.[12]

1.3 The synoptic problem and some proposed solutions

In New Testament studies, the synoptic problem is concerned with hypotheses about the relationships between the gospels according to Matthew, Mark and Luke. In what follows, we shall refer to these gospels simply as Matthew, Mark and Luke, and, on occasion, use the abbreviations Mt, Mk and Lk, respectively. The problem is how to explain the fact that there is so much similar material and yet also so many differences between Matthew, Mark and Luke, sometimes minor differences in wording and sometimes major differences, where longer sections of material are present in one gospel but absent in another, or present in a substantially different form.

The texts of the gospels may be partitioned into sections, commonly referred to as *pericopes* by New Testament scholars. Each such *pericope*[13] is a reasonably self-contained section of text, which may be a section of narrative material or a section of teaching, such as a parable, or a combination of both. Naturally, there are some differences of opinion as to how the text should be partitioned, but on the whole there seems to be broad agreement about the specification of most of the pericopes. Some of the pericopes are unique to just one of the synoptic gospels, and such material is known as *single tradition*. So, for example, the birth and infancy narratives of the first two chapters of Luke's gospel, including the familiar Christmas story of the birth of Jesus and of the appearance of the angel to the shepherds, are single tradition material. So too are the very different birth and infancy narratives of the first two chapters of Matthew, including the story of the visit of the wise men, the Magi. Mark, on the other hand, has no birth and infancy narrative at all. Other pericopes

[12]Nowadays, there are sophisticated computer packages available as an aid to biblical studies, whether in the original languages or in English translation, enabling a variety of searches and analyses of the texts to be carried out. I have made substantial use of the Accordance bible software (`www.accordancebible.com`).

[13]The term *pericope* comes from a Greek word meaning "a cutting around" or "a piece cut out".

are common to just two of the synoptic gospels, and they are known as *double tradition*. The details of the wording of such pericopes will nevertheless differ to a greater or lesser extent between the gospels. The pericopes that make up Matthew's famous Sermon on the Mount are predominantly double tradition, in that they are to be found also in the gospel of Luke but not in Mark. However, although they are gathered together in a single block of teaching in Matthew, they are scattered in different locations in Luke. The majority of double tradition pericopes are those that are common to Matthew and Luke, and usually, nowadays, the term *double tradition* is restricted to these, but there are smaller numbers of double tradition pericopes that are common to Mark and Matthew or to Mark and Luke, and we shall follow Honoré (1968) in using the term *double tradition* to include these as well. Finally, there is the *triple tradition* of pericopes that are common to all three synoptic gospels, which include a great variety of accounts of healings, miracles and the teaching of Jesus, and most of the passion narrative. In most cases it is straightforward to classify pericopes into single, double or triple tradition, but in some problematic cases, and especially where there is disagreement about the definition of the pericopes, the classification may be less obvious. There are, for example, a few cases of pericopes that may be regarded as consisting partly of triple tradition and partly of double tradition material.[14]

The various published synopses generally present the text pericope by pericope, but the layout of the gospel parallels may vary considerably between the synopses, partly because of variant specifications of the pericopes, but also because the order in which the pericopes appear varies from gospel to gospel. Because of this latter fact, different compilers of synopses may choose different orderings of the pericopes in their presentation of the material or have duplicate copies of the same pericope in separate locations. A helpful overview of the pericopes and how their locations vary between the gospels is provided by Barr (1995) in his *Diagram of Synoptic Relationships*.

Because of the complex patterns of similarities and dissimilarities between the synoptic gospels, the problem of how to account for the relationships between them is a notoriously difficult one. To what extent has any gospel writer used the gospels of his predecessors and what other sources, oral or written, may he have had? As little is known about the history of the early church in the second half of the first century, hypotheses about the relationships between the synoptic gospels are based almost entirely upon the internal evidence of the texts themselves. On the other hand, any decision on what synoptic hypothesis to adopt will have implications for our understanding of early church history. It will also have an impact on how we read and interpret the gospels, whether in the context of academic study or personal devotion.[15] Good introductions to the issues involved in the synoptic problem and the various models that have been proposed are provided by Stein (2001) and Goodacre (2001).

[14]For specific examples, see the pericopes listed in Table 6.11 on p. 121.

[15]For a very personal view of the matter, from the standpoint of a particular hypothesis, see Farmer (1994).

A more detailed historical perspective on the study of the synoptic problem is given by Dungan (1999).

Some of the textual parallels between the synoptic gospels are so close that some kind of literary relationship between them is indicated,[16] with later writers copying and editing the work of their predecessors, but in considering the differences between the texts of the synoptic gospels the role of oral tradition must also be borne in mind. It has long been accepted that in the early church oral traditions played an important role in the transmission of the material that came to be incorporated into the gospels. However, as pointed out by Dunn (2003a,b), discussion of the synoptic problem has come to be in terms of literary relationships, while the role of oral transmission has faded into the background. Dunn has now attempted to reverse this tendency by emphasising the essentially oral culture in which the gospel writers operated. The transmission of gospel material would have been through oral performance, where, on the whole, the performers or teachers would faithfully transmit core material about the life and, perhaps especially, the teaching of Jesus, but where there would be considerable flexibility and variation from performance to performance in the details of the presentation as it was adapted to different settings and audiences. As a result, variant traditions would emerge, having "a characteristic combination of stability and diversity, of fixity and flexibility", "the same yet different".[17] The gospel writers would have been immersed in a culture of such oral performance even if they also had some written sources available, and, as the gospels came to be written and started to circulate in document form, the primary means of transmission of Jesus traditions in what was predominantly a non-literate society would still have been through oral performance. Furthermore, we may envisage the emergence of a secondary layer of orality, where the written gospel texts influence the later oral tradition. Where there are considerable discrepancies among the texts of the synoptic gospels, this, according to Dunn, may be particularly suggestive of the influence of oral tradition. Bauckham (2006) has argued for the primacy of eyewitnesses in the transmission of the traditions that came to be incorporated into the gospels, and Eve (2013) provides a survey of recent work on the oral tradition and its relation to issues of individual and collective memory.

The extent to which we should envisage the oral tradition as influencing the gospel writers is debatable. As Eve emphasises, any discussion of the interrelationship between the oral tradition and the composition of the gospels is highly speculative. Nevertheless, he concludes that "There is nothing wrong with the suggestion that oral tradition may have influenced the Evangelists' use of material also available to them in a written source; indeed, given the

[16]See, for example, Farmer (1964), pp. 202-208, or Chapter 1 of Stein (2001).

[17]Dunn's articles in this area have been brought together in Dunn (2013). See especially pp. 280-281, 305-306. The summary phrase, "the same yet different", is a favourite of Dunn's and appears frequently in his articles.

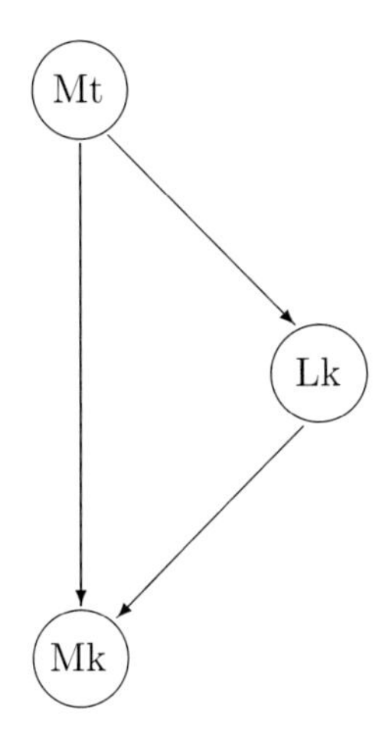

Figure 1.1
The Griesbach hypothesis

nature of the oral-scribal interface in antiquity this seems highly likely".[18] For further discussion of the interaction between orality and literacy and the danger of over-emphasising the role of the oral tradition, see Chapter 8 of Goodacre (2012) and the summary presented as Thesis III of Watson (2013).[19] In any case, the data that we have to work with are the written texts as they have come down to us, and these are the data on which we shall base our analyses. However, this is not at all to deny the influence of oral tradition, which will be discussed in the interpretation of the results, especially with regard to the examples of sections of text in Chapter 7.

At the dawn of the modern era of critical biblical scholarship, Johann Jakob Griesbach (1745-1812) has a particularly important place in the history of the synoptic problem. In 1776 he first published as a separate volume his Greek synopsis of the synoptic gospels. His contributions to New Testament studies and in particular to the analysis of the synoptic problem are celebrated in Orchard and Longstaff (1978). In this modern era, the first synoptic hypothesis to gain a large degree of acceptance among New Testament scholars was the *Griesbach hypothesis*, which was the dominant one in the late eighteenth and early nineteenth century. According to Griesbach, Matthew's was the first gospel to be written. Matthew was used by Luke, and Mark was a conflation of Matthew and Luke. This hypothesis is shown schematically in Figure 1.1.

[18]Eve (2013), p. 183
[19]Watson (2013), pp. 608-609

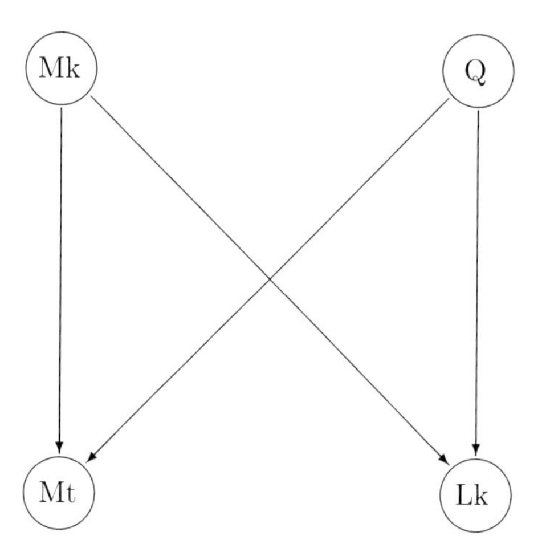

Figure 1.2
The two-source hypothesis

The Griesbach hypothesis has been revived in the twentieth century by Farmer (1964, 1994) and then further refined by McNicol et al. (1996) and Peabody et al. (2002), who prefer to refer to their reformulation as the "two-gospel hypothesis". On the other hand, challenges to the revival of the Griesbach hypothesis are mounted by Tuckett (1983) and Goodacre (1999).

In the nineteenth century there emerged the *two-source hypothesis* (also known as the *two-document hypothesis*), according to which Mark's was the first of the synoptic gospels to be written. Furthermore, Mark was used independently by Matthew and Luke, but they also independently of each other used another hypothetical source Q, which has not survived but which accounts for the large quantity of double tradition material common to Matthew and Luke but absent from Mark. This hypothesis is shown schematically in Figure 1.2. The two-source hypothesis became the dominant one and has remained so to the present day, so that textbooks often present it as more or less established fact. A classic treatment in the English language is that of Streeter (1924), who explicitly includes additional sources, M and L, for the single tradition material of Matthew and Luke, respectively, and hence uses the term *four-document hypothesis* for his extended model. More recent expositions of the two-source hypothesis and the state of Q research are provided by Kloppenborg (1987, 2000, 2008), Catchpole (1993) and Tuckett (1996). For all his emphasis on the role of oral transmission, Dunn (2003a,b) also supports the

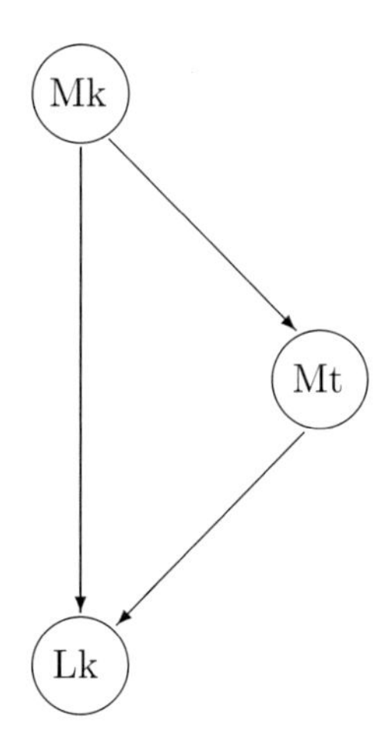

Figure 1.3
The Farrer hypothesis

two-source hypothesis and the existence of a written Q source.[20] Among those with a more thoroughgoing commitment to a Q source, there has emerged a scholarly industry devoted to reconstructing the lost text of Q, based on the Matthew-Luke double tradition, and even providing a historical and social setting for its development through a series of editions. An important product of this line of research is *The Critical Edition of Q* of Robinson et al. (2000). Crook (2012) in his synopsis presents an English version of this reconstructed Q in parallel with the texts of the gospels.

However, over the last few decades, a serious challenge has been mounted to the two-source hypothesis by the revival of the Griesbach hypothesis and especially by the emergence of what is known as the *Farrer hypothesis*, according to which Mark was the first synoptic gospel to be written, Matthew used Mark, and Luke used both Mark and Matthew, as illustrated in Figure 1.3. The seminal paper of Farrer (1955) was popularized and built upon by Goulder (1974, 1989). Currently the principal champion of the Farrer hypothesis is Goodacre (1996, 2002), but recently Watson (2009, 2013) has also come out in support.[21] For further explorations of the Farrer hypothesis see also the papers in Goodacre and Perrin (2004) and in Poirier and Peterson (2015, in press). Other models for synoptic relationships also receive support from time

[20]See, for instance, Dunn (2003a), p. 158, or Dunn (2003b), pp. 147f.

[21]See in particular Chapters 3 and 4 of Watson (2013). Instead of “Farrer hypothesis”, Watson prefers the terminology “*L/M* hypothesis” or “*L/M* theory”.

to time, and some of these will appear in the relevant later chapters. A useful selection of materials relating to the debate about the two-source hypothesis may be found in Bellinzoni (1985). From the nineteenth century onwards, some variants of the two-source hypothesis have supposed that Matthew and Luke used not Mark as we now have it, but an earlier version or versions of Mark — Ur-Mark or Proto-Mark. Such an approach has been taken in recent years by Burkett (2004, 2009), but the inclusion of a postulated but no longer extant Proto-Mark does not fall within the scope of the models that we shall consider in this study.

The major issue that lies behind the statistical analyses in this book is whether it is necessary to postulate the existence of a written Q source, as is implicit in the two-source hypothesis, or whether synoptic relationships are better modelled without recourse to Q.

1.4 Statistics, verbal agreements and sources of data

1.4.1 Stylometry

The use of statistical data in the analysis of the synoptic problem goes back to at least the end of the nineteenth century, when Hawkins' *Horae Synopticae* was published, the first edition in 1899. The second edition, Hawkins (1909), is a still very useful handbook for students of the synoptic problem, whose title ("Synoptic Hours") points to the innumerable hours that the author spent poring over the texts of the gospels. It contains a wealth of data about the synoptic gospels, including statistics of word frequencies to demonstrate what words and phrases are particularly characteristic of each of the gospel writers.

Nowadays the term *stylometry* is used to refer to the analysis of the relative frequencies of commonly occurring words, or types of word, or their positions in sentences, and other statistical properties of texts, such as the distribution of sentence length, to characterise the writing style of particular authors. Where the authorship of a text is unknown or disputed, its stylometric characteristics may be used in an attempt to identify the author. Discussion of such methods together with applications may be found, for example, in Morton (1965, 1978), Michaelson and Morton (1972), Kenny (1986), Holmes (1998), Stamatatos (2009) and in a statistically more rigorous presentation in Mosteller and Wallace (1984). Bartholomew (1988, 1996), in his wide-ranging reviews of the uses of probability and statistics in theology and biblical studies, has included sections on stylometry. In New Testament studies the main question to be investigated by such methods has been which of the 13 epistles that purport to be written by the apostle Paul are genuinely Pauline. Multivariate statistical methods have been used on this problem by Neumann (1990), Ledger (1995) and Mealand (1996). More recently, Mealand (2011) has adopted a stylometric approach and used more sophisticated multivariate statistical techniques specifically for the synoptic problem, in order to investigate the existence of the Q source.[22]

1.4.2 Verbal agreements

Some of the arguments about the relative merits of the various hypotheses about synoptic relationships have been based upon the differences in order of the pericopes in the three synoptic gospels. The image is often used of the pericopes as beads, which have been strung together on a string, in a different order in each gospel. But arguments from order appear to be inconclusive, as has been reaffirmed by Neville (2002), who carried out a thorough examination of the arguments, including some statistical analysis.

In this book, however, we shall be concerned neither with stylometric techniques, nor with arguments from order, but with the analysis of verbal agreements, an approach that goes back to the pioneering paper of Honoré (1968). A *verbal agreement* between any pair of gospels, A and B, occurs where the same word is present in the same context in both gospels. So if B is using A as a source then the word in question has been retained unchanged by B as opposed to being omitted or altered, although the exact position of the word may be different through reordering of the words or by the omission or interpolation of other words. The Koine Greek used in the gospels is a highly inflected language, and, following Honoré, by a verbal agreement we shall mean that the word is present in both gospels in the same grammatical form. It should be noted that in this approach, in common with stylometry, the basic units of analysis are words, and in recording verbal agreements the meaning of the words is not taken into account, only their form. (The meaning of the words is, however, essential to the construction of the pericopes that provide the context for verbal comparisons.) To see examples of how verbal agreements arise and may be recorded, Figure 7.2 and subsequent figures in Chapter 7 may be consulted together with the key to the underlinings provided in Figure 7.1.

It has been a matter of debate as to what definition of verbal agreement should be used. Others, such as Carlston and Norlin (1971), have used less restrictive definitions of verbal agreement than did Honoré, and the issues involved have been discussed by O'Rourke (1974) and Matilla (1994). Carlston and Norlin (1999) later conceded that with hindsight they might have used a tighter definition of verbal agreement, yet one that was nevertheless broader than Honoré's. They also showed that, even if they had used Honoré's definition, their conclusions regarding the two-source model would have been the same. The history of the statistical analysis of verbal agreements for the synoptic problem has been comprehensively reviewed by Poirier (2008).

Honoré (1968) constructed his own data set and his paper is especially

[22]For a recent introduction to computational stylometry and related topics, including examples from New Testament studies, see Oakes (2014).

useful in that it provides a listing of the data that were used in the analysis, including counts of verbal agreements, and a detailed account of his mathematical and statistical reasoning. Since then, more detailed data on verbal agreements have been published, by Morgenthaler (1971) and Tyson and Longstaff (1978). We shall be making particular use of the latter, whose definition of verbal agreements corresponds very closely to Honoré's. Nevertheless, where comparisons are made, the counts of verbal agreements in Honoré (1968) and Tyson and Longstaff (1978) do differ somewhat. This occurs for a number of reasons:

1. There are slight differences in the detailed specification of what is counted as a verbal agreement.
2. Differing arrangements of the text into pericopes are used, which can affect the context used for identifying verbal agreements.
3. Occasionally a piece of text that appears in one pericope in one of the gospels appears in a totally different pericope in another gospel. Differing judgments may be made as to whether any verbal agreements are then to be registered.
4. Particular problems arise when *doublets* occur, two different versions of a piece of text in different locations in the same gospel. Different decisions can then be made as to what is the appropriate context for counting verbal agreements.
5. Different versions of the Greek text are used. Honoré (1968) used the text of Huck (1949), but Tyson and Longstaff (1978) used Aland (1971).
6. Last, but certainly not least, human error can and does occur in the identification, counting and recording of verbal agreements.

Overall, then, there is a certain amount of subjectivity and error involved in the determination of verbal agreements, which has some effect on the recorded counts. The evidence in Chapter 2 suggests that the resulting differences in the recorded data do not have any serious effect on the conclusions to be drawn from the analysis.

In the later chapters of this book we shall make use of data sets on verbal agreements that we have constructed from the *Synopticon* produced by Farmer (1969). Using the 25th edition of the standard Nestle-Aland text,[23] Farmer presented the Greek text of each of the synoptic gospels with individual words highlighted in different colours to indicate for which of them there was verbal agreement with words in each of the other two synoptic gospels. In the case of Mark's gospel, words that appear unchanged in Matthew only are highlighted in yellow, words that appear unchanged in Luke only are highlighted in green, and words that appear unchanged in both Matthew and Luke are highlighted

[23] Nestle-Aland (1963), NA[25]

in blue. As will be described in Chapter 3, this colour-coding for Mark has been translated into a bivariate binary time series. The data set that we shall use in Chapter 6 is similarly constructed from Farmer's colour-coding for Matthew.

1.5 Outline of this book

In Chapter 2, we build upon the triple-link model introduced by Honoré (1968), broadly following the development in Abakuks (2006a,b, 2007), but with some changes. In Chapter 3, the bivariate binary time series obtained from the highlighted text of Mark in Farmer's *Synopticon* is introduced, and much of the rest of the book is concerned with the analysis of such binary time series. In Chapter 3 itself, following the treatment in Abakuks (2012), under the assumption of Markan priority, i.e., that Mark was the first of the synoptic gospels to be written, the time series is analyzed using logistic regression methods in order to investigate Matthew's and Luke's use of Mark and, in particular, to test whether Matthew and Luke were statistically independent in their verbal agreements with Mark. From Chapter 4 onwards, hidden Markov models are used to analyze the series. Chapter 4 is a theoretical interlude in which the theory of hidden Markov models for binary data is presented, with some associated R code in Appendix A at the end of the book. In Chapter 5, hidden Markov models are used to investigate Matthew's and Luke's use of Mark.[24] In Chapter 6, a different bivariate binary time series, which uses Matthew as the base text instead of Mark, is analysed in order to investigate the verbal agreements of Mark and Luke with Matthew.

After the statistical analysis in earlier chapters of the binary data extracted from the gospel texts, in Chapter 7 we turn to some examples of the incomparably richer material of the texts themselves — in Greek, to be able to exhibit the verbal agreements word by word, and in English translation, to make the texts accessible to a wider readership. Parallel passages from the synoptic gospels, which emerge from the results of the statistical analysis as particularly significant for the synoptic problem, are presented for detailed examination. We focus especially on passages that, from the statistical analysis, appear likely to provide the strongest evidence that, under the assumption of Markan priority, Matthew and Luke were not independent in their use of Mark. This leads to the discussion of specific issues concerning the relationships between parallel sections of text and the responses of New Testament scholars who defend different synoptic hypotheses. Finally, in Chapter 8 we summarize the conclusions that may be drawn from our analysis and suggest directions for further research.

To put things in the broader context of New Testament studies, much of

[24] An introductory sketch of this approach is given in Abakuks (2015, in press).

the material in this book could be regarded as falling within the scope of what is known as *source criticism*, which seeks to identify the sources used by the gospel authors. Some of the discussion in Chapter 7 touches on aspects of *redaction criticism*, which deals with the way in which the gospel authors edited and adapted their sources to fit in with their own theological standpoint. These and other aspects of biblical criticism are surveyed in Tuckett (1987).

2

The triple-link model

2.1 Introduction

Honoré (1968) introduced the terminology of *double-link model* and *triple-link model* to describe families of models for synoptic relationships. In what follows, like Honoré, we use the terms Gospel A, B and C to refer to any permutation of the synoptic gospels, Matthew (Mt), Mark (Mk) and Luke (Lk).

The double-link models are those that in the mathematical terminology of graph theory are represented by a directed graph with three nodes and two directed edges or arrows, as shown in Figure 2.1. The nodes represent the three gospels and the arrows each represent the direction from source to user. In what Honoré called the "linear model", Gospel B uses Gospel A as a source, and Gospel C uses Gospel B, but there is no direct use of Gospel A by Gospel C. In the "fork model", Gospels A and C both make use of Gospel B as a source, but Gospel C does not make use of Gospel A, or vice versa. In what we may refer to as the "conflation model", Gospel B uses both Gospel A and Gospel C as sources.

In the triple-link model the associated graph has three arrows, as illustrated in Figure 2.2. It is assumed that Gospels B and C both use Gospel A as a source and that Gospel C also uses Gospel B. For later use the diagram also illustrates the notation x for the proportion of words in A that are transmitted unchanged to B, y for the proportion of words in B that are transmitted unchanged to C, and z for the proportion of words in A that are transmitted unchanged directly to C, where the concept of a word being transmitted unchanged is defined in terms of verbal agreement as discussed in Section 1.4.

When all the permutations of A, B and C (or Mt, Mk and Lk) are taken into account, there are 6 different cases of the linear model, 3 of the fork model, 3 of the conflation model and 6 of the triple-link model, a total of 18 in all, which were individually listed by Farmer (1964), pp. 208, 209. Farmer, however, readily concluded that the double-link models were insufficient to account for the patterns of agreements between the gospels, which left only the 6 cases of the triple-link model to be considered. In fact the cases of the triple-link model do correspond to synoptic models that have been proposed and received varying levels of support, notably the ones that we have already described in Chapter 1. For instance, the Griesbach model (see Figure 1.1),

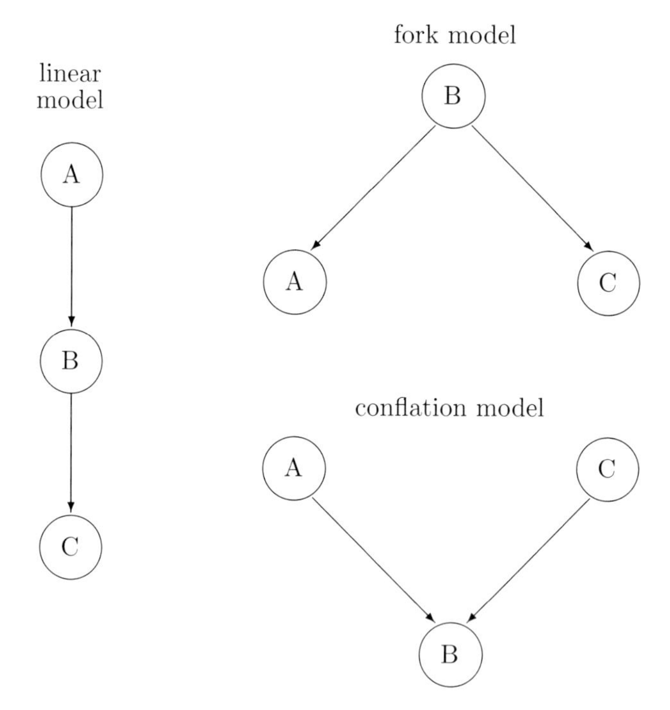

Figure 2.1
Double-link models

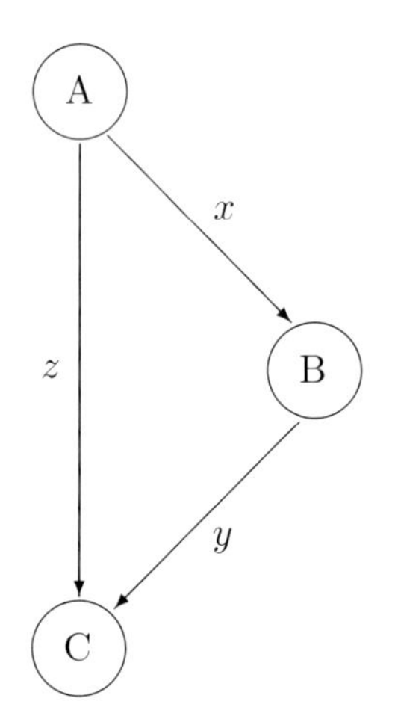

Figure 2.2
The triple-link model

favoured by Farmer, corresponds to the case A = Mt, B = Lk, C = Mk. The Farrer model (see Figure 1.3) corresponds to the case A = Mk, B = Mt, C = Lk. Butler (1939, 1951) and Wenham (1991) have supported what is often referred to as the Augustinian hypothesis that corresponds to the triple-link model with A = Mt, B = Mk, C = Lk, according to which the gospels were written in the order that they appear in the *New Testament*, Matthew, followed by Mark, followed by Luke.[1] The triple-link model with A = Mk, B = Lk, C = Mt has been supported by Huggins (1992) and Hengel (2000). However, the cases of the triple-link model with A = Lk, according to which the Gospel of Luke was the first to be written, have received little support.[2]

It should also be noted that, although a number of well-supported synoptic hypotheses are encompassed by the triple-link model, the two-source hypothesis (see Figure 1.2), which is the synoptic hypothesis that probably still has the greatest following, is not. In this chapter we shall focus mainly on the analysis of the triple-link model, but in Section 2.5 we shall also look at the double-link models, which, although they are not adequate models in their own right, may still be thought of as building blocks for more complex models.

It is a commonly made observation that all statistical models are simplifications which provide only approximations to the situation that they are attempting to represent. Kloppenborg (2000) makes a number of related points specifically with regard to models for synoptic relationships.[3] Thus: "Synoptic hypotheses are simplifications ... unlikely to represent precisely or fully the *actual* compositional processes of the gospels. ... Hypotheses are heuristic models intended to aid comprehension and discovery; they do not replicate reality". The double-link and triple-link models are simplifications that aim to represent the literary relationships between the three synoptic gospels. They do not explicitly take into account other possible literary sources or the influence of oral tradition, but neither do they rule them out.

2.2 The data

Honoré (1968) classified the pericopes (or sections, as he called them) into triple tradition, double tradition and single tradition and tabulated counts of verbal agreements by pericope, together with totals aggregated over the triple

[1] Peabody (1983) has observed that Augustine of Hippo (354-430), in addition to the hypothesis named after him, later came also to espouse another view of the synoptic gospels according to which Mark drew on both Matthew and Luke, so anticipating the Griesbach hypothesis. For a discussion of Augustine's views on the relationships between the gospels see also Chapter 1 of Watson (2013).

[2] A dissenting view about the writing of the synoptic gospels that dates them to around 140 AD and gives priority to Luke may be found in Vinzent (2011, 2014).

[3] See Kloppenborg (2000), pp. 50-54, for his discussion of the issues.

tradition, double tradition and single tradition, respectively. In his analysis of the double-link model and the triple-link model, Honoré used, firstly, the data from the triple tradition and, secondly, the data from the whole of the synoptic material, stating that[4] "for various reasons it is simplest to concentrate on the triple tradition material in the first instance", without being any more specific. It is certainly the case that this is the collection of pericopes where there is apparent interdependence between all three synoptic gospels. We shall also make use of the data from the triple tradition and double tradition combined, that is, the whole of the synoptic material less the single tradition. This latter set of data includes all the material where there appear to be some links between the synoptic gospels, but excludes blocks of material which are unique to any gospel author, i.e., material where the author may have had his own special source and which has not been taken up by any subsequent author or, possibly, material from a common hypothetical source, available to all or two of the evangelists, but used by only one of them. This is arguably the most natural set of data to use, but we may reserve judgment on which is the most appropriate data set until we have completed the analysis. As we shall see, the triple tradition and double tradition combined does provide the best support for the triple-link model. Table 2.1 gives the counts of words classified according to their presence or absence in each of the synoptic gospels,

(a) for the triple tradition,

(b) for the triple and double tradition combined, and

(c) for the whole of the synoptic material.

We shall use all three sets of data to examine how well the triple-link model fits the data. The data are taken from Honoré (1968), Tables 1, 2, 4 and 10, although presented here in a different way. The first three columns of Table 2.1 are indicator variables that represent the presence or absence of words in the respective gospels, so that, in any row of the table, the counts refer to the number of words that are present in the gospels marked with the number 1 but absent from the gospels marked with the number 0.

For example, focusing on the whole of the text of Matthew, there are, according to our definition of verbal agreement and the corresponding counts in the final column of Table 2.1, 1852 words that Matthew has in common with both Mark and Luke, 2735 words that Matthew has in common with Mark but not with Luke, 2386 words that Matthew has in common with Luke but not with Mark, and 11292 words that are unique to Matthew, giving a total of

$$1852 + 2735 + 2386 + 11292 = 18265$$

words in Matthew.

Following Farmer (1969), Tyson and Longstaff (1978), pp. 9-11, worked

[4]Honoré (1968), p. 97

Table 2.1
Honoré's counts of verbal agreements in the synoptic gospels

Indicators			Triple	Triple	All synoptic
Mt	Mk	Lk	tradition	plus double	material
1	1	1	1852	1852	1852
1	1	0	1908	2735	2735
1	0	1	637	2386	2386
0	1	1	1039	1165	1165
0	0	1	4356	7231	13957
0	1	0	3831	5269	5576
1	0	0	3939	7588	11292

Table 2.2
Tyson & Longstaff's counts of verbal agreements in the synoptic gospels

Indicators			Triple	Triple	All synoptic
Mt	Mk	Lk	tradition	plus double	material
1	1	1	1732	1732	1732
1	1	0	2270	2700	2700
1	0	1	894	2239	2239
0	1	1	1074	1141	1141
0	0	1	5313	7564	14264
0	1	0	4495	5320	5452
1	0	0	4799	8520	11622

with three definitions of verbal agreement, one of which corresponds very closely to Honoré's. Using this definition, the data in Tyson and Longstaff (1978), pp. 169-171, enable us to construct Table 2.2 that has the same form as Table 2.1 for Honoré's data. The numbers in the two tables are somewhat different for the various reasons listed in Section 1.4. Tyson and Longstaff have a more detailed classification of pericopes that includes the concept of a pericope being "partially parallel" in a pair of gospels. In constructing our counts for triple tradition and double tradition, we have made no distinction between a pericope that is present in any pair of gospels and one that is only partially parallel.

Using a stricter definition of verbal agreement, Tyson and Longstaff (1978), also provided, in their terminology, counts of "identical words in continuous agreement". These are counts of words that are identical in two comparable passages and that are immediately preceded or followed by words that are identical, i.e., they are part of a block of identical words. This then amounts to a more stringent definition of agreement that takes into account the order in

Table 2.3
Tyson & Longstaff's counts of identical words in continuous agreement

Indicators			Triple	Triple	All synoptic
Mt	Mk	Lk	tradition	plus double	material
1	1	1	1140	1140	1140
1	1	0	2050	2372	2372
1	0	1	773	1846	1846
0	1	1	937	995	995
0	0	1	6163	8695	15395
0	1	0	5444	6386	6518
1	0	0	5732	9833	12935

Table 2.4
Counts of verbal agreements with Mark (Honoré)

		Luke		
		0	1	total
Matthew	0	5576	1165	6741
	1	2735	1852	4587
	total	8311	3017	11328

which words occur and that leads to smaller counts of agreements — isolated words that are in identical agreement are not counted as agreements according to this definition. Table 2.3 is of the same form as Table 2.1 and Table 2.2, but uses this stricter definition of verbal agreement.

We shall later use data from all these three tables to investigate the plausibility of the triple-link model, but for the present, to provide some initial overview, we extract from Table 2.1 and Table 2.2 contingency tables of the counts of verbal agreements of Matthew and Luke with Mark: Table 2.4 uses the final column of Honoré's data in Table 2.1, and Table 2.5 uses the final column of Tyson and Longstaff's data in Table 2.2. In both Table 2.4 and Table 2.5 the value 1 of the indicator variable in a row or column indicates counts of verbal agreements with Mark and the value 0 indicates counts of non-agreements.

It is natural to use Mark as the base text for this first look at the data, because the most widely supported synoptic models assume that Mark was the first of the synoptic gospels to be written and that it was used as a source

Table 2.5
Counts of verbal agreements with Mark (Tyson & Longstaff)

		Luke		
		0	1	total
Matthew	0	5452	1141	6593
	1	2700	1732	4432
	total	8152	2873	11025

Table 2.6
The lengths in words of the synoptic gospels

	Honoré	Tyson & Longstaff
Matthew	18265	18293
Mark	11328	11025
Luke	19360	19376

by both Matthew and Luke. In Chapters 3 and 5 too we shall use Mark as the base text.

Again we see that there are differences between Honoré's data in Table 2.4 and Tyson and Longstaff's data in Table 2.5. According to Honoré's data, there are 11328 words in Mark, of which, under the assumption of Markan priority, 1852 are retained unchanged by both Matthew and Luke, but according to Tyson and Longstaff's data there are 11025 words in Mark, of which 1732 are retained unchanged by both Matthew and Luke. Part of the discrepancy arises because Honoré has included the longer ending of Mark's gospel, Mk 16:9-20, a total of 167 words, which nowadays is generally regarded as a later addition to the gospel, whereas Tyson and Longstaff have not included it in their counts. Removal of this section of Markan single tradition, would result in a reduction of 167 of the count in the top left-hand cell in Table 2.4 and in the same reduction in the corresponding row and column totals and the grand total, so that there would be a total of 11161 words counted in Mark.

Using the data of Table 2.4 as it stands, and assuming that Matthew and Luke used Mark as a source, we may note that Matthew follows Mark more closely, retaining $4587/11328 = 40\%$ of Mark's words unchanged, whereas Luke retains $3017/11328 = 27\%$ of Mark's words unchanged. (Using the data of Table 2.5, the corresponding percentages are 40% and 26%, respectively.)

It is also worth noting at this stage that Matthew's and Luke's gospels are considerably longer than Mark's gospel, as shown in Table 2.6.

2.3 Probabilistic notation and assumptions

Although the situation has now been somewhat redressed by Poirier (2008), what little attention the statistical analysis in Honoré (1968) has received in the biblical studies literature[5] has been almost entirely restricted to his investigation of the two-source model and the issue of the existence of Q, but a considerable part of Honoré's paper dealt with an innovative analysis of his triple-link model. O'Rourke (1974) was dismissive of the triple-link model on the grounds that it gave a very poor fit to the observed data, but his conclusions were based on Honoré's analysis, which turns out to be flawed. In this chapter we shall show that it is possible to find a good fit.

Honoré (1968) was careful to give a detailed account of his assumptions and reasoning in the fitting of the triple-link model, but, for a professional statistician, one of the challenges of Honoré's paper is that his terminology tends not to conform to what is accepted usage in statistical theory. In this section we recast Honoré's analysis in the language and notation of mathematical probability theory. This helps to make more explicit certain assumptions that he made and to provide a more rigorous confirmation of some of his results. It also helps to show where Honoré went astray and to come up with a more satisfactory analysis.

In the triple-link model, as illustrated in Figure 2.2, the symbols x, y and z refer to proportions of words that are transmitted from one gospel to another, but now we envisage them as probabilities:

— x the probability that a given word in Gospel A is transmitted unchanged to Gospel B,

— y the probability that a given word in Gospel B is transmitted unchanged to Gospel C,

— z the probability that a given word in Gospel A is transmitted unchanged directly to Gospel C.

In order to analyse the data using the notation of probability theory, we shall define the events A, B and C:

— A the event that a given word is in Gospel A,

— B the event that a given word is in Gospel B,

— C the event that a given word is in Gospel C.

We shall further introduce events C_1 and C_2:

— C_1 the event that a given word is in Gospel C and has been transmitted to Gospel C via Gospel B,

[5] See Carlston and Norlin (1971, 1999), O'Rourke (1974), Matilla (1994, 2004).

— C_2 the event that a given word is in Gospel C and has been transmitted to Gospel C directly from Gospel A.

According to the triple-link model, any word that Gospel C has in common with either Gospel A or Gospel B has been transmitted to Gospel C from either Gospel A or Gospel B, neglecting the possibility that it might have been transmitted to Gospel C from some other unspecified source. It follows that

$$A \cap C = A \cap (C_1 \cup C_2),$$
$$B \cap C = B \cap (C_1 \cup C_2),$$
$$A \cap B \cap C = A \cap B \cap (C_1 \cup C_2).$$

We also note that the occurrence of the event C_1 implies B, so that we have $C_1 \subseteq B$ and hence $C_1 \cap \bar{B} = \emptyset$, where $\bar{B}$ denotes the complementary event that B does not occur and $\emptyset$ denotes the empty set. Similarly, the occurrence of the event C_2 implies A, so that $C_2 \subseteq A$ and hence $C_2 \cap \bar{A} = \emptyset$. It follows that

$$\bar{A} \cap B \cap C = \bar{A} \cap B \cap C_1 \tag{2.1}$$

and

$$A \cap \bar{B} \cap C = A \cap \bar{B} \cap C_2. \tag{2.2}$$

Equation (2.1) expresses the statement that a word that is not in Gospel A, but is in Gospels B and C, has been transmitted to Gospel C from Gospel B. Similarly, Equation (2.2) expresses the statement that a word that is not in Gospel B, but is in Gospels A and C, has been transmitted to Gospel C directly from Gospel A.

With the same notation for events as introduced in the previous paragraph, $\Pr(B|A)$ denotes the conditional probability that a given word is in Gospel B given that it is in Gospel A. Using the basic definition of conditional probability,

$$\Pr(B|A) = \frac{\Pr(A \cap B)}{\Pr(A)},$$

we shall evaluate this conditional probability directly from the data by the corresponding relative frequency, that is, the ratio of the number of words that are in both Gospels A and B to the number of words that are in Gospel A. The conditional probability so evaluated is precisely the probability that, for the part of the synoptic material currently under consideration, a word chosen at random from Gospel A is also in Gospel B — in the same setting and in the same grammatical form. Similar direct evaluations can be made for all conditional probabilities involving A, B and C, but conditional probabilities that involve C_1 and C_2 will have to be evaluated indirectly.

In terms of the notation that we have introduced, the probabilities x, y

and z may be expressed as

$$x = \Pr(B|A), \tag{2.3}$$
$$y = \Pr(C_1|B), \tag{2.4}$$
$$z = \Pr(C_2|A). \tag{2.5}$$

We shall want to evaluate the probabilities x, y and z, which we shall do by expressing them as conditional probabilities which can be evaluated directly. It is straightforward to evaluate x directly from Equation (2.3), but expressions for y and z are not so immediate. In fact Honoré (1968) in calculating his Table 6 went astray by working as if y was given by $\Pr(C|B)$. In practice, all calculated values of x, y and z lie strictly between 0 and 1, which we shall implicitly assume where necessary in the following theoretical development.

In our analysis, we shall follow Honoré in making certain conditional independence assumptions.

Assumption 1 – given that a word is in Gospel A, the event that it is transmitted to Gospel B and the event that it is transmitted directly from Gospel A to Gospel C are independent. Thus

$$\Pr(B \cap C_2|A) = \Pr(B|A)\Pr(C_2|A),$$

which is equivalent to

$$\Pr(C_2|A \cap B) = \Pr(C_2|A).$$

Assumption 2 – given that a word is in Gospel B, the event that it is in Gospel A and the event that it is transmitted from Gospel B to Gospel C are independent. Thus

$$\Pr(A \cap C_1|B) = \Pr(A|B)\Pr(C_1|B),$$

which is equivalent to

$$\Pr(C_1|A \cap B) = \Pr(C_1|B).$$

Assumption 3 – given that a word is in Gospel A and Gospel B, the event that it is transmitted from Gospel B to Gospel C and the event that it is transmitted directly from Gospel A to Gospel C are independent. Thus

$$\Pr(C_1 \cap C_2|A \cap B) = \Pr(C_1|A \cap B)\Pr(C_2|A \cap B).$$

Using Assumptions 1 and 2 it follows from Assumption 3 that

$$\Pr(C_1 \cap C_2|A \cap B) = \Pr(C_1|B)\Pr(C_2|A). \tag{2.6}$$

Using his own terminology of word selections being "unbiased", Honoré stated his use of Assumptions 1 and 2, while not acknowledging his implicit use of Assumption 3. As pointed out by Honoré (1968), p. 101, such assumptions will by no means hold exactly. For example, with regard to Assumption 1 and assuming Markan priority (A = Mk), although it might reasonably be assumed that Matthew and Luke used the text of Mark independently of each other, in the sense of not collaborating or not being directly influenced by each other in their use of Mark, this does not imply that they would be statistically independent in the choice of words that they retained unchanged directly from Mark. Rather, we might well expect the editorial criteria that Matthew and Luke used to select words directly from Mark to have some common features, leading to some statistical dependence. This will become apparent from the detailed examination of the examples of text in Chapter 7. However, it should be noted that in the triple-link model a word that is in Gospel A can be transmitted unchanged to Gospel C not only directly but also indirectly via Gospel B. In this way, even under Assumption 1, Gospel B exerts an influence on the use of Gospel A in Gospel C, which we shall demonstrate explicitly and discuss more fully in Section 2.4.1, specifically in the comments on the inequality (2.15).

Honoré's assumptions were also criticized by Wenham (1972), pp. 14-15. With regard to the Griesbach hypothesis, for example, for which A = Mt, B = Lk, C = Mk, if Mark was attempting to produce a conflation of Matthew and Luke, he would be more likely to use Lukan material that Luke had taken from Matthew as against other Lukan material. This, according to Wenham, contradicts Assumption 2. Wenham's point would pose a serious problem for the triple-link model if Assumption 2 implied that $\Pr(C|A \cap B) = \Pr(C|\bar{A} \cap B)$, but it does not. As we shall see explicitly in the inequality (2.14) of Section 2.4.1, $\Pr(C|A \cap B) > \Pr(C|\bar{A} \cap B)$. A word that is present in Gospel A and in Gospel B can arrive in Gospel C in two ways, either via B or directly from A, so that a word that is in A and B is more likely to be present also in C than a word that is present in B but not in A.

When we consider Honoré's implicit Assumption 3 together with ways in which we might imagine modeling the mental processes and compositional techniques of the author of Gospel C, the following Assumption 3A emerges as a possible alternative to Assumption 3. If we use Assumptions 1 and 2 together with Assumption 3A, we shall refer to this model as the *modified triple-link model*.

Assumption 3A – the event C_1 that a word is transmitted from Gospel B to Gospel C and the event C_2 that it is transmitted directly from Gospel A to Gospel C are mutually exclusive, i.e., $C_1 \cap C_2 = \emptyset$.

It follows, for example, from Assumption 3A that

$$\Pr(C_1 \cup C_2|A \cap B) = \Pr(C_1|A \cap B) + \Pr(C_2|A \cap B)$$

and hence, using Assumptions 1 and 2, that

$$\Pr(C_1 \cup C_2|A \cap B) = \Pr(C_1|B) + \Pr(C_2|A). \tag{2.7}$$

In trying to assess which of the Assumptions 3 and 3A is the more plausible, and bearing in mind that any model is a simplification of the underlying process of composition of the gospels, it is nevertheless useful to consider the physical conditions under which the gospel writers worked. Much relevant background material is reviewed in Chapter 5 of Neville (2002). This is an area in which Downing has published a number of papers, many of which are collected together in Downing (2000), although it should be noted that he is a staunch defender of the two-source hypothesis as against any of the hypotheses that underlie the triple-link model. However, some of the most important points for our purposes are already to be found in the much earlier work of Sanday (1911), pp. 16-19. Although discussion of the conditions under which a gospel writer worked can only be highly speculative, we should certainly not imagine him as sitting in a study with tables or desks on which the earlier texts were immediately available for reference. In the first century, the gospels were written on scrolls, which were expensive, scarce and, furthermore, difficult to work with, requiring some effort to move from one section to another. It is unlikely that the gospel writer would have been looking at two or more scrolls simultaneously. It is more likely that, if on any occasion he was reading a section of a scroll as his primary source, or having it read to him, he would be relying on his memory to compare this with any other material, whether written or oral, that was known to him. For a discussion of this see in particular Downing (1988, 1992), but for a more recent study that suggests the use of wax tablets as a compositional aid see Poirier (2012).

So the author of Gospel C might, perhaps, at any one time be looking either at a scroll of Gospel A or a scroll of Gospel B, but not both, and possibly, from time to time, might switch from one to the other. In such a scenario, the assumption that C_1 and C_2 are mutually exclusive would seem to be more natural than the assumption that they are conditionally independent. On the other hand, if the author of Gospel C was using a scroll of Gospel B, say, and working from memory with the text of Gospel A, then it does not seem obvious whether the assumption of conditional independence or of mutual exclusivity better models the process of composition. In that case, we may simply try fitting both the original triple-link model and the modified version and see which appears to fit best.

2.4 A probabilistic analysis

Given our data, as remarked in Section 2.3, we shall be able to evaluate x directly from the expression of Equation (2.3). In this section, using the assumptions of Section 2.3, we obtain expressions that will allow us to evaluate z and y and to examine how well the model fits the data. The expression for z is obtained using Assumption 1 and the expression for y is obtained using Assumption 2, which are the same for both Honoré's original model and the modified model that was proposed in Section 2.3. Thereafter, the analysis depends on which version of the model is chosen, and we present the subsequent analysis in two sub-sections, the first for Honoré's original model, using Assumption 3, and the second for the modified model, using Assumption 3A.

Using Equation (2.2) and the definition of conditional probability,

$$\Pr(C|A \cap \bar{B}) = \Pr(C_2|A \cap \bar{B}).$$

In the formulae of Assumption 1 we may validly replace B by $\bar{B}$. Hence

$$\Pr(C|A \cap \bar{B}) = \Pr(C_2|A \cap \bar{B}) = \Pr(C_2|A) = z,$$

using the definition of z in Equation (2.5). Thus

$$z = \Pr(C|A \cap \bar{B}) = \frac{\Pr(A \cap \bar{B} \cap C)}{\Pr(A \cap \bar{B})}, \tag{2.8}$$

which will enable us to evaluate z.

Using Equation (2.1) and the definition of conditional probability,

$$\Pr(C|\bar{A} \cap B) = \Pr(C_1|\bar{A} \cap B).$$

In the formulae of Assumption 2 we may validly replace A by $\bar{A}$. Hence

$$\Pr(C|\bar{A} \cap B) = \Pr(C_1|\bar{A} \cap B) = \Pr(C_1|B) = y,$$

using the definition of y in Equation (2.4). Thus

$$y = \Pr(C|\bar{A} \cap B) = \frac{\Pr(\bar{A} \cap B \cap C)}{\Pr(\bar{A} \cap B)}, \tag{2.9}$$

which will enable us to evaluate y.

It should be noted that in Equation (2.9) we have adopted a simpler, more direct formula for evaluating y than the formulae that were used in Abakuks (2006a) for the original triple-link model and in Abakuks (2007) for the modified model. In these papers the formulae used for evaluating y, which follow from our Equations (2.11) and (2.17) that are obtained in sub-sections 2.4.1 and 2.4.2, may be thought of as adjustments of Honoré's erroneous use of $\Pr(C|B)$. The Equations (2.8) and (2.9) that we shall now use for evaluating z and y, respectively, are similar in form. In the corresponding ratios of word counts that are used for carrying out the calculations, words that are common to Gospel A and Gospel B are in both cases excluded. Intuitively, this is because any such word that is also in Gospel C could have been taken either directly from Gospel A or from Gospel B, and we cannot identify which is the case.

2.4.1 The original triple-link model

We first obtain, for the original model, expressions for a number of conditional probabilities and then comment on some of their implications and uses. Using the Assumptions 1, 2 and 3, and specifically Equation (2.6), we obtain

$$\begin{aligned}
\Pr(C|A\cap B) &= \Pr(C_1\cup C_2|A\cap B)\\
&= \Pr(C_1|A\cap B)+\Pr(C_2|A\cap B)-\Pr(C_1\cap C_2|A\cap B)\\
&= \Pr(C_1|B)+\Pr(C_2|A)-\Pr(C_1|B)\Pr(C_2|A)\\
&= y+z-yz,
\end{aligned}$$

using the definitions of Equations (2.4) and (2.5). Thus

$$\Pr(C|A\cap B)=y+(1-y)z. \tag{2.10}$$

Hence, using Equations (2.10) and (2.9),

$$\begin{aligned}
\Pr(C|B) &= \Pr(C|A\cap B)\Pr(A|B)+\Pr(C|\bar{A}\cap B)\Pr(\bar{A}|B)\\
&= [y+(1-y)z]\Pr(A|B)+y[1-\Pr(A|B)]\\
&= y+(1-y)z\Pr(A|B). \qquad (2.11)
\end{aligned}$$

Using Equations (2.3) and (2.10) we find that

$$\begin{aligned}
\Pr(B\cap C|A) &= \Pr(C|A\cap B)\Pr(B|A)\\
&= xy+xz-xyz. \qquad (2.12)
\end{aligned}$$

Furthermore,

$$\begin{aligned}
\Pr(C|A) &= \Pr(B\cap C|A)+\Pr(\bar{B}\cap C|A)\\
&= \Pr(B\cap C|A)+\Pr(\bar{B}\cap C_2|A)\\
&= \Pr(B\cap C|A)+\Pr(\bar{B}|A)\Pr(C_2|A),
\end{aligned}$$

using Assumption 1. Substituting in from Equations (2.3), (2.5) and (2.12),

$$\begin{aligned}
\Pr(C|A) &= xy+xz-xyz+(1-x)z\\
&= z+xy-xyz. \qquad (2.13)
\end{aligned}$$

From Equation (2.11) it follows that $\Pr(C|B)>y$, so that we see explicitly that it is inappropriate, as was done by Honoré (1968), to evaluate y as if it were given by $\Pr(C|B)$.

From Equations (2.9) and (2.10) we see that

$$\Pr(C|A\cap B)=\Pr(C|\bar{A}\cap B)+(1-y)z,$$

from which it follows that

$$\Pr(C|A\cap B)>\Pr(C|\bar{A}\cap B). \tag{2.14}$$

To guard against any possible misinterpretation of Assumption 2, we note that it follows from the inequality (2.14) that a word that is present in Gospel B is more likely to be present in Gospel C if it is also present in Gospel A than if it is not also present in Gospel A.

Starting with Equation (2.12) and then using Equations (2.3) and (2.13) we find that

$$\begin{aligned}\Pr(B \cap C|A) &= x(y + z - yz)\\ &= \Pr(B|A)[z + y(1 - z)]\\ &> \Pr(B|A)[z + xy(1 - z)]\\ &= \Pr(B|A)\Pr(C|A).\end{aligned}$$

Thus

$$\Pr(B \cap C|A) > \Pr(B|A)\Pr(C|A),$$

or equivalently

$$\Pr(C|A \cap B) > \Pr(C|A),$$

so that the events B and C are not conditionally independent given A. This fact has to be distinguished from what is stated in Assumption 1, that the events B and C_2 are conditionally independent given A. What the above inequalities express is the fact that there is a positive association between the words from Gospel A that are retained unchanged in Gospel B and Gospel C. To put it in slightly different terms, from Equations (2.8) and (2.10),

$$\Pr(C|A \cap B) = \Pr(C|A \cap \bar{B}) + y(1 - z),$$

so that

$$\Pr(C|A \cap B) > \Pr(C|A \cap \bar{B}). \tag{2.15}$$

According to the inequality (2.15), a word from Gospel A is more likely to be retained unchanged in Gospel C if it has been retained unchanged in Gospel B than if it has not been retained unchanged in Gospel B. In that sense, Gospel B exerts an influence on the use of Gospel A by Gospel C. This is a theme that will appear again and again with regard to the examples of text and the associated data presented in Chapter 7.

The expressions of Equations (2.12) and (2.13) are identical with those obtained by Honoré (1968), p. 104, without explicit use of the probability calculus. Honoré used these expressions in his discussion of the validity of the triple-link model. We shall use our values for x, y and z to calculate $\Pr(C|A \cap B)$ and $\Pr(C|A)$ using Equations (2.10) and (2.13), respectively, and then attempt to assess the fit of the triple-link model to the data by checking how close these calculated values are to the direct evaluations obtained from ratios of observed word counts. The use of Equation (2.10) for $\Pr(C|A \cap B)$ is for this purpose equivalent to the use of Equation (2.12) for $\Pr(B \cap C|A)$, which was the equation used in Abakuks (2006a).

2.4.2 The modified triple-link model

We repeat the analysis of the previous sub-section, but now for the modified model, and obtain parallel, but somewhat simpler results. Using the Assumptions 1, 2 and 3A, and specifically Equation (2.7), we obtain

$$\begin{aligned}\Pr(C|A\cap B) &= \Pr(C_1\cup C_2|A\cap B)\\ &= \Pr(C_1|B)+\Pr(C_2|A)\\ &= y+z,\end{aligned}\tag{2.16}$$

using the definitions of Equations (2.4) and (2.5). Hence, using Equations (2.16) and (2.9),

$$\begin{aligned}\Pr(C|B) &= \Pr(C|A\cap B)\Pr(A|B)+\Pr(C|\bar{A}\cap B)\Pr(\bar{A}|B)\\ &= (y+z)\Pr(A|B)+y[1-\Pr(A|B)]\\ &= y+z\Pr(A|B).\end{aligned}\tag{2.17}$$

Using Equations (2.3) and (2.16) we find that

$$\begin{aligned}\Pr(B\cap C|A) &= \Pr(C|A\cap B)\Pr(B|A)\\ &= x(y+z).\end{aligned}\tag{2.18}$$

Furthermore, just as for the original model,

$$\begin{aligned}\Pr(C|A) &= \Pr(B\cap C|A)+\Pr(\bar{B}\cap C|A)\\ &= \Pr(B\cap C|A)+\Pr(\bar{B}\cap C_2|A)\\ &= \Pr(B\cap C|A)+\Pr(\bar{B}|A)\Pr(C_2|A).\end{aligned}$$

Now substituting in from Equations (2.3), (2.5) and (2.18),

$$\begin{aligned}\Pr(C|A) &= xy+xz+(1-x)z\\ &= z+xy.\end{aligned}\tag{2.19}$$

We can obtain very similar conclusions to the ones drawn for the original model. From Equation (2.17) it follows that $\Pr(C|B)>y$. From Equations (2.9) and (2.16) we see that

$$\Pr(C|A\cap B)=\Pr(C|\bar{A}\cap B)+z,$$

so that

$$\Pr(C|A\cap B)>\Pr(C|\bar{A}\cap B),$$

which is identical with the inequality (2.14) in Section 2.4.1.

From Equations (2.18), (2.3) and (2.19) we find that

$$\Pr(B\cap C|A)>\Pr(B|A)\Pr(C|A),$$

so that again there is a positive association between the words from Gospel

A that are retained unchanged in Gospel B and Gospel C. From Equations (2.8) and (2.16),

$$\Pr(C|A \cap B) = \Pr(C|A \cap \bar{B}) + y,$$

so that

$$\Pr(C|A \cap B) > \Pr(C|A \cap \bar{B}),$$

which is identical with the inequality (2.15) in Section 2.4.1.

We shall use Equations (2.16) and (2.19) to assess the fit of the modified triple-link model. The expressions in these equations are simpler than the corresponding expressions in Equations (2.10) and (2.13) for the original model, which both included an additional term,$-yz$ and $-xyz$, respectively, on the right-hand side.

2.5 A statistical analysis

We present the results of the calculations based on the probabilistic analysis of Section 2.4 and using the data of Section 2.2. The calculations are for the six possible cases of the triple-link model for both the original triple-link model and the modified triple-link model, using in turn the data for the triple tradition, the triple plus double tradition combined, and the whole of the synoptic material. The data from Table 2.1 are used to produce Table 2.7 for Honoré's data, the data from Table 2.2 are used to produce Table 2.8 for Tyson and Longstaff's data on verbal agreements, and the data from Table 2.3 are used to produce Table 2.9 for Tyson and Longstaff's data on identical words in continuous agreement.

We illustrate the calculations for the first row of Table 2.7, which uses Honoré's data for the triple tradition from Table 2.1. Taking A = Matthew, B = Mark, C = Luke, and evaluating the conditional probability as the ratio of the corresponding word frequencies, we have from Equation (2.3),

$$x = \Pr(B|A) = \frac{\Pr(A \cap B)}{\Pr(A)} = \frac{1852 + 1908}{1852 + 1908 + 637 + 3939} = 0.451.$$

From Equation (2.9),

$$y = \Pr(C|\bar{A} \cap B) = \frac{\Pr(\bar{A} \cap B \cap C)}{\Pr(\bar{A} \cap B)} = \frac{1039}{1039 + 3831} = 0.213.$$

From Equation (2.8),

$$z = \Pr(C|A \cap \bar{B}) = \frac{\Pr(A \cap \bar{B} \cap C)}{\Pr(A \cap \bar{B})} = \frac{637}{637 + 3939} = 0.139.$$

Next the value of $\Pr(C|A \cap B)$ is calculated in three different ways and the results compared. Firstly a direct calculation is made,

$$\Pr(C|A \cap B) = \frac{\Pr(A \cap B \cap C)}{\Pr(A \cap B)} = \frac{1852}{1852 + 1908} = 0.493.$$

For the original triple-link model, from Equation (2.10),[6]

$$\Pr(C|A \cap B) = y + (1 - y)z = 0.213 + (0.787)(0.139) = 0.323.$$

How well these two evaluations of $\Pr(C|A \cap B)$ agree is measured by their ratio,

$$\frac{0.323}{0.493} = 0.655.$$

For the modified triple-link model, from Equation (2.16),

$$\Pr(C|A \cap B) = y + z = 0.213 + 0.139 = 0.353.$$

How well this evaluation agrees with the direct calculation is measured by their ratio,

$$\frac{0.353}{0.493} = 0.716.$$

Finally the value of $\Pr(C|A)$ is calculated in three different ways and the results compared. Firstly a direct calculation is made,

$$\Pr(C|A) = \frac{\Pr(A \cap C)}{\Pr(A)} = \frac{1852 + 637}{1852 + 1908 + 637 + 3939} = 0.299.$$

For the original triple-link model, from Equation (2.13),

$$\begin{aligned} \Pr(C|A) &= z + xy - xyz \\ &= 0.139 + (0.451)(0.213) - (0.451)(0.213)(0.139) = 0.222. \end{aligned}$$

How well these two evaluations of $\Pr(C|A)$ agree is measured by their ratio,

$$\frac{0.222}{0.299} = 0.744.$$

For the modified triple-link model, from Equation (2.19),

$$\Pr(C|A) = z + xy = 0.139 + (0.451)(0.213) = 0.235.$$

How well this evaluation agrees with the direct calculation is measured by their ratio,

$$\frac{0.235}{0.299} = 0.789.$$

[6]The numbers here and in the following illustrative calculations are presented to 3 decimal places, but the calculations are carried out to a higher degree of accuracy, so that results are correct to 3 decimal places, with no rounding errors.

Table 2.7
Fitting the triple-link model to Honoré's data

Triple tradition material:				Calculations for $\Pr(C\|A \cap B)$					Calculations for $\Pr(C\|A)$				
					original		modified			original		modified	
A-B-C	x	y	z	direct	$y+z-yz$	ratio	$y+z$	ratio	direct	$z+xy-xyz$	ratio	$z+xy$	ratio
Mt-Mk-Lk	0.451	0.213	0.139	0.493	0.323	0.655	0.353	0.716	0.299	0.222	0.744	0.235	0.789
Lk-Mk-Mt	0.367	0.332	0.128	0.641	0.418	0.652	0.460	0.718	0.316	0.234	0.741	0.249	0.790
Mk-Mt-Lk	0.436	0.139	0.213	0.493	0.323	0.655	0.353	0.716	0.335	0.261	0.779	0.274	0.818
Lk-Mt-Mk	0.316	0.326	0.193	0.744	0.456	0.613	0.519	0.697	0.367	0.276	0.752	0.296	0.806
Mt-Lk-Mk	0.299	0.193	0.326	0.744	0.456	0.613	0.519	0.697	0.451	0.365	0.809	0.384	0.851
Mk-Lk-Mt	0.335	0.128	0.332	0.641	0.418	0.652	0.460	0.718	0.436	0.361	0.829	0.375	0.861

Triple + double tradition material:				Calculations for $\Pr(C\|A \cap B)$					Calculations for $\Pr(C\|A)$				
					original		modified			original		modified	
A-B-C	x	y	z	direct	$y+z-yz$	ratio	$y+z$	ratio	direct	$z+xy-xyz$	ratio	$z+xy$	ratio
Mt-Mk-Lk	0.315	0.181	0.239	0.404	0.377	0.934	0.420	1.041	0.291	0.283	0.971	0.296	1.018
Lk-Mk-Mt	0.239	0.342	0.248	0.614	0.505	0.823	0.590	0.961	0.335	0.309	0.923	0.330	0.983
Mk-Mt-Lk	0.416	0.239	0.181	0.404	0.377	0.934	0.420	1.041	0.274	0.263	0.959	0.281	1.025
Lk-Mt-Mk	0.335	0.265	0.139	0.437	0.367	0.840	0.404	0.924	0.239	0.215	0.902	0.228	0.953
Mt-Lk-Mk	0.291	0.139	0.265	0.437	0.367	0.840	0.404	0.924	0.315	0.295	0.935	0.305	0.969
Mk-Lk-Mt	0.274	0.248	0.342	0.614	0.505	0.823	0.590	0.961	0.416	0.386	0.928	0.410	0.984

All the synoptic material:				Calculations for $\Pr(C\|A \cap B)$					Calculations for $\Pr(C\|A)$				
					original		modified			original		modified	
A-B-C	x	y	z	direct	$y+z-yz$	ratio	$y+z$	ratio	direct	$z+xy-xyz$	ratio	$z+xy$	ratio
Mt-Mk-Lk	0.251	0.173	0.174	0.404	0.317	0.785	0.347	0.860	0.232	0.210	0.906	0.218	0.939
Lk-Mk-Mt	0.156	0.329	0.146	0.614	0.427	0.696	0.475	0.774	0.219	0.190	0.867	0.197	0.901
Mk-Mt-Lk	0.405	0.174	0.173	0.404	0.317	0.785	0.347	0.860	0.266	0.231	0.868	0.243	0.914
Lk-Mt-Mk	0.219	0.195	0.077	0.437	0.257	0.588	0.272	0.622	0.156	0.116	0.747	0.120	0.768
Mt-Lk-Mk	0.232	0.077	0.195	0.437	0.257	0.588	0.272	0.622	0.251	0.209	0.834	0.213	0.848
Mk-Lk-Mt	0.266	0.146	0.329	0.614	0.427	0.696	0.475	0.774	0.405	0.355	0.877	0.368	0.909

Table 2.8
Fitting the triple-link model to Tyson & Longstaff's data

Triple tradition material:				Calculations for $\Pr(C\|A\cap B)$					Calculations for $\Pr(C\|A)$				
					original		modified			original		modified	
A-B-C	x	y	z	direct	$y+z-yz$	ratio	$y+z$	ratio	direct	$z+xy-xyz$	ratio	$z+xy$	ratio
Mt-Mk-Lk	0.413	0.193	0.157	0.433	0.320	0.738	0.350	0.808	0.271	0.224	0.828	0.237	0.874
Lk-Mk-Mt	0.311	0.336	0.144	0.617	0.431	0.699	0.480	0.777	0.291	0.233	0.801	0.248	0.853
Mk-Mt-Lk	0.418	0.157	0.193	0.433	0.320	0.738	0.350	0.808	0.293	0.246	0.839	0.259	0.882
Lk-Mt-Mk	0.291	0.321	0.168	0.660	0.435	0.660	0.489	0.742	0.311	0.246	0.790	0.262	0.841
Mt-Lk-Mk	0.271	0.168	0.321	0.660	0.435	0.660	0.489	0.742	0.413	0.352	0.853	0.367	0.888
Mk-Lk-Mt	0.293	0.144	0.336	0.617	0.431	0.699	0.480	0.777	0.418	0.364	0.870	0.418	0.903

Triple + double tradition material:				Calculations for $\Pr(C\|A\cap B)$					Calculations for $\Pr(C\|A)$				
					original		modified			original		modified	
A-B-C	x	y	z	direct	$y+z-yz$	ratio	$y+z$	ratio	direct	$z+xy-xyz$	ratio	$z+xy$	ratio
Mt-Mk-Lk	0.292	0.177	0.208	0.391	0.348	0.890	0.385	0.984	0.261	0.249	0.952	0.260	0.993
Lk-Mk-Mt	0.227	0.337	0.228	0.603	0.488	0.810	0.565	0.937	0.313	0.287	0.917	0.305	0.973
Mk-Mt-Lk	0.407	0.208	0.177	0.391	0.348	0.890	0.385	0.984	0.264	0.246	0.934	0.261	0.991
Lk-Mt-Mk	0.313	0.241	0.131	0.436	0.340	0.780	0.372	0.852	0.227	0.197	0.867	0.206	0.911
Mt-Lk-Mk	0.261	0.131	0.241	0.436	0.340	0.780	0.372	0.852	0.292	0.267	0.914	0.275	0.942
Mk-Lk-Mt	0.264	0.228	0.337	0.603	0.488	0.810	0.565	0.937	0.407	0.377	0.926	0.397	0.975

All the synoptic material:				Calculations for $\Pr(C\|A\cap B)$					Calculations for $\Pr(C\|A)$				
					original		modified			original		modified	
A-B-C	x	y	z	direct	$y+z-yz$	ratio	$y+z$	ratio	direct	$z+xy-xyz$	ratio	$z+xy$	ratio
Mt-Mk-Lk	0.242	0.173	0.162	0.391	0.307	0.785	0.335	0.856	0.217	0.197	0.906	0.203	0.937
Lk-Mk-Mt	0.148	0.331	0.136	0.603	0.422	0.700	0.467	0.774	0.205	0.178	0.869	0.185	0.902
Mk-Mt-Lk	0.402	0.162	0.173	0.391	0.307	0.785	0.335	0.856	0.261	0.227	0.870	0.238	0.913
Lk-Mt-Mk	0.205	0.189	0.074	0.436	0.249	0.570	0.263	0.602	0.148	0.110	0.741	0.113	0.760
Mt-Lk-Mk	0.217	0.074	0.189	0.436	0.249	0.570	0.263	0.602	0.242	0.202	0.832	0.205	0.844
Mk-Lk-Mt	0.261	0.136	0.331	0.603	0.422	0.700	0.467	0.774	0.402	0.355	0.883	0.367	0.912

Table 2.9
Fitting the triple-link model to Tyson & Longstaff's data for identical words in continuous agreement

Triple tradition material:				Calculations for $\Pr(C\|A \cap B)$					Calculations for $\Pr(C\|A)$				
					original		modified			original		modified	
A-B-C	x	y	z	direct	$y+z-yz$	ratio	$y+z$	ratio	direct	$z+xy-xyz$	ratio	$z+xy$	ratio
Mt-Mk-Lk	0.329	0.147	0.119	0.357	0.248	0.695	0.266	0.743	0.197	0.161	0.818	0.167	0.847
Lk-Mk-Mt	0.230	0.274	0.111	0.549	0.355	0.646	0.385	0.701	0.212	0.167	0.789	0.174	0.822
Mk-Mt-Lk	0.333	0.119	0.147	0.357	0.248	0.695	0.266	0.743	0.217	0.181	0.832	0.186	0.859
Lk-Mt-Mk	0.212	0.263	0.132	0.596	0.361	0.605	0.395	0.664	0.230	0.181	0.783	0.188	0.815
Mt-Lk-Mk	0.197	0.132	0.263	0.596	0.361	0.605	0.395	0.664	0.329	0.283	0.859	0.289	0.880
Mk-Lk-Mt	0.217	0.111	0.274	0.549	0.355	0.646	0.385	0.701	0.333	0.291	0.873	0.298	0.893

Triple + double tradition material:				Calculations for $\Pr(C\|A \cap B)$					Calculations for $\Pr(C\|A)$				
					original		modified			original		modified	
A-B-C	x	y	z	direct	$y+z-yz$	ratio	$y+z$	ratio	direct	$z+xy-xyz$	ratio	$z+xy$	ratio
Mt-Mk-Lk	0.231	0.135	0.158	0.325	0.272	0.837	0.293	0.902	0.197	0.184	0.938	0.189	0.963
Lk-Mk-Mt	0.168	0.271	0.175	0.534	0.399	0.746	0.446	0.835	0.236	0.213	0.903	0.221	0.937
Mk-Mt-Lk	0.322	0.158	0.135	0.325	0.272	0.837	0.293	0.902	0.196	0.179	0.913	0.186	0.948
Lk-Mt-Mk	0.236	0.194	0.103	0.382	0.277	0.726	0.297	0.778	0.168	0.144	0.854	0.148	0.881
Mt-Lk-Mk	0.197	0.103	0.194	0.382	0.277	0.726	0.297	0.778	0.231	0.211	0.911	0.215	0.928
Mk-Lk-Mt	0.196	0.175	0.271	0.534	0.399	0.746	0.446	0.835	0.322	0.296	0.918	0.305	0.947

All the synoptic material:				Calculations for $\Pr(C\|A \cap B)$					Calculations for $\Pr(C\|A)$				
					original		modified			original		modified	
A-B-C	x	y	z	direct	$y+z-yz$	ratio	$y+z$	ratio	direct	$z+xy-xyz$	ratio	$z+xy$	ratio
Mt-Mk-Lk	0.192	0.132	0.125	0.325	0.241	0.742	0.257	0.793	0.163	0.147	0.901	0.150	0.921
Lk-Mk-Mt	0.110	0.267	0.107	0.534	0.345	0.647	0.374	0.700	0.154	0.133	0.865	0.136	0.886
Mk-Mt-Lk	0.319	0.125	0.132	0.325	0.241	0.742	0.257	0.793	0.194	0.167	0.862	0.172	0.889
Lk-Mt-Mk	0.154	0.155	0.061	0.382	0.206	0.540	0.216	0.565	0.110	0.083	0.755	0.085	0.768
Mt-Lk-Mk	0.163	0.061	0.155	0.382	0.206	0.540	0.216	0.565	0.192	0.163	0.851	0.165	0.859
Mk-Lk-Mt	0.194	0.107	0.267	0.534	0.345	0.647	0.374	0.700	0.319	0.282	0.885	0.288	0.903

The ratios of the conditional probability values as calculated from the model formulae to the values calculated directly from the counts of verbal agreements may be regarded as measures of how well the triple-link model, either in its original or its modified form, fits the data — for each of the six possible cases of the model, for each of the three sets of material analysed, and for each of the three data sets presented in Tables 2.7, 2.8 and 2.9, respectively. The closer the value of the ratio to 1, the better the fit of the model. It is to be expected that the ratio will be less than 1 in all or almost all cases, especially for the original form of the model, since it is an acknowledged shortcoming of the model that somewhat unrealistic assumptions of statistical independence are made which will have the effect of making the predicted number of words that the gospels have in common smaller than what is actually observed. If that turns out to be the case then the larger the ratio the better the model has performed. The expression of Equation (2.16) for $\Pr(C|A \cap B)$, for the modified model, necessarily gives a greater value than the expression of Equation (2.10) for the original model. Similarly, the expression of Equation (2.19) for $\Pr(C|A)$, for the modified model, necessarily gives a greater value than the expression of Equation (2.13) for the original model, so that the modified model will in each case give a greater value of the ratio. Hence we might anticipate that the modified model will generally give a better fit than the original model. In fact there are just four places, in Table 2.7, where for the modified version of the model the ratio is slightly greater than 1, in which case the greater ratio does not necessarily correspond to the better fit.

When we take an overview of Tables 2.7, 2.8 and 2.9, we observe that the model fits best when the triple plus double tradition material is used, which, as suggested in Section 2.2, is arguably the most natural version of the data to use. The modified version of the triple-link model does indeed give a better fit than the original version. We shall limit our discussion here to the triple plus double tradition data, for which Honoré's data of Table 2.7 give a better fit than Tyson and Longstaff's data of Table 2.8, which in turn give a better fit than the data in Table 2.9 that are based on the more restrictive definition of identical words in continuous agreement.

When we compare the goodness of fit of the six different cases of the model, we should note first of all that the ratios for $\Pr(C|A \cap B)$ are the same for each pair of cases that have the same Gospel C, i.e., the same gospel as the final one to be written. If we adopt the better fitting modified model, it is clear that the cases with C = Mk are the ones that perform worst according to this ratio criterion. For the Tyson and Longstaff data in both Table 2.8 and Table 2.9, the cases with C = Lk perform best. For Honoré's data in Table 2.7, the situation is less clear-cut in that the ratio 1.041 for C = Lk now exceeds 1 and the ratio 0.961 for C = Mt is about as far below 1 as the ratio for C = Lk is above 1. However, for Honoré's original model the cases with C = Lk clearly perform best.

Turning to the ratios for $\Pr(C|A)$, we find that we can now discriminate between all six cases of the model. Comparison of the ratios here broadly conforms to what was found for the ratios for $\Pr(C|A \cap B)$, but we can go somewhat further. Of the cases with C = Lk, the Mt-Mk-Lk model, which corresponds to the Augustinian hypothesis, gives a slightly better fit than the Mk-Mt-Lk model, which corresponds to the Farrer hypothesis. Of the cases with C = Mt, the Mk-Lk-Mt model gives a slightly better fit than the Lk-Mk-Mt model. The Lk-Mt-Mk model gives the uniformly worst fit.

To come up with another criterion for the validity of a specific case of the triple-link model, Honoré (1968), pp. 118-119, argued that, if the gospels were written in the order A-B-C then, since B has only A as his source, whereas C has both A and B as sources, one would expect C to make less use of each of A and B than B does of A. Furthermore, C should be expected to make more use of his most recent source B than of the earlier source A. Thus Honoré expects (i) C's use of B to be greater than C's use of A and (ii) B's use of A to be greater than C's use of B. In terms of our notation, Honoré's criterion appears as

$$x > y > z, \tag{2.20}$$

although he himself did not express it in this way. But Honoré's criterion is open to criticism. We may envisage a scenario in which C has long been familiar with A, but the more recent B then comes into C's hands, after which he writes his own version. In such a situation we might expect C to make more use of the long familiar A than of B, while still accepting with Honoré that C makes less use of each of A and B than B does of A. This translates into the inequalities

$$x > z > y. \tag{2.21}$$

If the conditions (2.20) and (2.21) are regarded as alternative possibilities, then the weaker condition

$$x > \max(y, z) \tag{2.22}$$

emerges. But there is no logical necessity for any of these conditions to be satisfied, since B could simply be following his source material less closely than is C. So in fact, x, y and z could be in any numerical order, and it would be dangerous to place much or any reliance on the use as criteria of any of the conditions expressed in the inequalities (2.20) – (2.22). Nevertheless, consideration of the numerical order of x, y and z for any fitted case of the triple-link model will be of interest as a way of comparing the extent to which gospel authors make use of their sources.

Using the triple plus double tradition data, of the two cases of the model which generally gave the best fit, the marginally better fitting Mt-Mk-Lk (Augustinian) model satisfies the inequalities (2.21), so that Mark follows Matthew more closely than does Luke, and Luke follows Matthew more closely than he does Mark. The Mk-Mt-Lk (Farrer) model satisfies the inequalities (2.20) of Honoré's criterion, so that Matthew follows Mark more closely than Luke follows Matthew, and Luke follows Matthew more closely than he does Mark.

As may be seen from comparison of the frequencies for the triple plus double tradition data in Tables 2.1, 2.2 and 2.3, Honoré finds somewhat greater numbers of verbal agreements than do Tyson and Longstaff, and the numbers of agreements are necessarily even smaller for the more stringent definition of identical words in continuous agreement. Correspondingly, for the triple plus double tradition data, the calculated values of x, y and z decrease uniformly from Table 2.7 to Table 2.8, and then again to Table 2.9. However, for each of the six cases of the model, the values of x, y and z remain in the same order, whichever of these tables we use, and hence our comments in the last paragraph remain the same.

2.6 Double-link models

Although the double-link models, illustrated in Figure 2.1, are in themselves inadequate for modelling the complexity of the relationships between the synoptic gospels, they are worth exploring at least briefly because they may be thought of as components of more realistic models and because they shed some light on certain modelling issues. In particular, the fork model with B=Mk, so that Mark is a source for Matthew and Luke, is a component of the two-source hypothesis illustrated in Figure 1.2.

Like Honoré (1968), we shall not treat the conflation model, which provides no explanation of words that are common to Gospel A and Gospel C, but restrict attention to the linear model and the fork model. In contrast to the triple-link model, there is in a double-link model only a single path by which a word can arrive in Gospel C, so that we may consider the event C that a given word is present in Gospel C without any need to consider events C_1 and C_2 as was necessary for the triple-link model.

For the linear model, following Honoré, we make an assumption that is essentially the same as Assumption 2 for the triple-link model: given that a word is in Gospel B, the event that it is in Gospel A and the event that it is in Gospel C are independent. Thus, with the events A, B and C as defined in Section 2.3,

$$\Pr(A \cap C|B) = \Pr(A|B)\Pr(C|B).$$

For the fork model, we make an assumption that is essentially the same as Assumption 1 for the triple-link model, with a change in the labelling of the gospels: given that a word is in Gospel B, the event that it is transmitted to Gospel A and the event that it is transmitted to Gospel C are independent. Thus again we have

$$\Pr(A \cap C|B) = \Pr(A|B)\Pr(C|B).$$

So for both these double-link models we make the same assumption that the events A and C are conditionally independent given the event B. As a consequence, the two models cannot be distinguished from each other in terms of the joint probability structure of A, B and C and any statistical analysis based upon that structure: the fork model with Gospel B as the source is in effect identical with the linear model with Gospel B as the middle term. In either case,

$$\frac{\Pr(A|B)\Pr(C|B)}{\Pr(A\cap C|B)} = 1.$$

In the fork model, the roles of Gospels A and C are interchangeable, so that there are just three distinct models: those with Matthew, Mark and Luke, respectively, as the source for the other two. Correspondingly, in terms of the joint probability structure, there are just three distinguishable linear models: those with Matthew, Mark and Luke, respectively, as the middle term. We may note that it is a consequence of the conditional independence assumption that in the linear model the roles of the first Gospel A and the last Gospel C are in effect interchangeable.

The fact that under our assumptions the linear model is indistinguishable from the fork model is related to the issue of what is known as the *Lachmann fallacy*.[7] The essence of the fallacious argument is as follows: if, according to some measure or measures, Gospels A and C are both more closely related to Gospel B than they are to each other then Gospel B must be a common source for Gospels A and C, i.e., a fork model with Gospel B as source is inferred. However, the relationship can equally be explained in terms of a linear model with Gospel B as the middle term.

It should be emphasised that such issues of model identifiability do not occur for the triple-link model, where there are six genuinely distinguishable cases corresponding to the six permutations of A, B and C.

To test the fit of the double-link models for the three distinguishable cases, Honoré (1968), followed by Abakuks (2006a), evaluated the ratio

$$\frac{\Pr(A|B)\Pr(C|B)}{\Pr(A\cap C|B)},$$

from the data of Table 2.1 using calculations of the form illustrated in Section 2.5 for the triple-link model. As might be expected, because of the questionable independence assumption, the ratios are substantially less than 1 in value and the double-link model does not appear to be able to provide an adequate fit to the data.

Another way of examining the assumption of conditional independence — one that is natural in terms of elementary statistical methods and also one that points forward to issues that will be addressed in later chapters — is to look at appropriate contingency tables. For the important case where Mark is taken to be the source, Table 2.4 for the Honoré data or Table 2.5 for the Tyson

[7]The Lachmann fallacy is discussed in Chapter 5 of Butler (1951) and in Palmer (1967). In fact, the argument used by the philologist Karl Lachmann (1793-1851) was not in itself fallacious, but it was misapplied by later authors.

Table 2.10
Honoré's counts of verbal agreements with Mark and, in brackets, expected numbers under the assumption of statistical independence

		Luke		
		0	1	total
Matthew	0	5576 (4946)	1165 (1795)	6741
	1	2735 (3365)	1852 (1222)	4587
	total	8311	3017	11328

and Longstaff data may be used to compare the observed frequencies with the expected frequencies under the assumption of statistical independence. In Table 2.10 we take Honoré's data from Table 2.4 and in brackets add, to the nearest integer, the corresponding expected frequencies for the given row and column totals.

We see immediately that along the diagonal of the contingency table the observed frequencies substantially exceed the expected frequencies, so that Matthew and Luke agree in what words from Mark to retain unchanged and what not to retain much more often than would be expected under the hypothesis of statistical independence. Thus there appears to be a positive association between Matthew and Luke in the words retained unchanged from Mark, which is consistent with what we would expect from a triple-link model in either its original or modified form, with A = Mark, as discussed in Sections 2.3 and 2.4.1, but not consistent with a double-link model. However, we have to be very cautious about what we may deduce with regard to the significance of these results, because the words of Mark cannot be regarded as a random sample whose members behave independently of each other. Words tend to be transmitted unchanged from one gospel to another in clusters of varying sizes, and there are large segments of material that are not transmitted at all. (A notable example is Luke's "great omission", where he appears to have made no use of the section of Mark's text from Mk 6:45-8:10.) We may calculate the usual chi-square statistic that is used to test the hypothesis of independence, but there is no justification for using the chi-square distribution with 1 degree of freedom to assess its significance. How to deal with this issue will be a major theme of the following chapters.

2.7 Conclusions

In this chapter we have made more explicit the assumptions that underlay Honoré's analysis of the triple-link model and have improved upon his statistical analysis of the model in order to see whether it may be used to provide a good fit to data on verbal agreements. We have also modified his implicit Assumption 3 to provide a modified version of the triple-link model.

As has been noted, the conditional independence assumptions are open to serious criticism. Despite this, we have found that, especially for the word counts from the triple plus double tradition material and for the modified version of the model, what appears to be a very good fit may be found, which overcomes the objection of O'Rourke (1974) that the model simply did not fit the observed data. Specifically, we may conclude that the cases of the triple-link model that identify Luke as the last of the synoptic gospels to have been written provide the best fit, that is, the case Mt-Mk-Lk which corresponds to the Augustinian hypothesis and the case Mk-Mt-Lk which corresponds to the Farrer hypothesis. As far as this analysis takes us, we have found models that fit the data well and do not require the existence of a Q source.

It is reassuring that our conclusions as to which cases of the models provide the best fit are fairly robust as to whether we use Honoré's data in Table 2.7, or Tyson and Longstaff's data in Table 2.8, or even the data for the more restrictive definition of continuous agreement in Table 2.9. The only noteworthy difference is that, for Honoré's data in conjunction with the modified model, the cases Lk-Mk-Mt and Mk-Lk-Mt, where Matthew is the last gospel to be written, come into contention as providing as good a fit as the cases for which Luke is the last. The least good fit is provided by the cases for which Mark is the last gospel to be written. Reassuringly, too, our conclusions are robust as to whether we calculate y as in this chapter or whether we use the formulae and tabulated results of Abakuks (2006a, 2007).

We may wish to go one step further and use our analysis to compare the plausibility of different models of synoptic relationships, for example, the Griesbach hypothesis (Mt-Lk-Mk) and the Farrer hypothesis (Mk-Mt-Lk), that lie behind the triple-link model. But here we have to be very careful, for these underlying models, although specified by linkages of the type illustrated in Figure 2.2, do not in themselves include the further assumptions, Assumptions 1, 2, 3/3A of Section 2.3, that we have adopted for our analysis and the resulting calculations of the ratios of Tables 2.7, 2.8 and 2.9. Consequently, when comparing the ratios, two issues are confounded: the extent to which any underlying synoptic model is valid, and the extent to which the three assumptions, and specifically the conditional independence assumptions, hold. So when, from examination of the ratios, it appears that our analysis gives more support to the Farrer hypothesis, it may be argued rather that the Griesbach hypothesis requires more substantial departures from the conditional independence assumptions. Nevertheless, a ranking based on the analysis of the 6 cases of the triple-link model, whether in its original or its modified form, should carry some weight in discussions of the relative merits of the corresponding synoptic hypotheses.

Another serious issue with regard to our method of analysis is that the results are presented essentially at the level of descriptive statistics, with no standard errors or any indication of how significant are the differences in the goodness of fit between the various cases of the triple-link model. We might well consider the transmission probabilities x, y and z as model parameters to be estimated, using appropriate estimators, and expect to see associated confidence intervals. Given the observed counts of verbal agreements, we could conceive of a likelihood function being written down for x, y and z, from which, using standard methods of analysis, techniques of statistical inference would be derived. A huge obstacle to carrying out such a programme is that, as already noted in the discussion of Table 2.10, the individual words cannot even remotely be regarded as behaving independently of each other. Because of this, there is no obvious way of writing down a likelihood function. Nevertheless, potentially there is scope for the development of a statistically more rigorous method of analysis of our data. As a first step in this direction, assuming Markan priority, in the following chapters we shall investigate in detail how we might model the transmission of the sequence of words from Mark to Matthew and Luke. In Chapter 6 we shall carry out a similar exercise assuming Matthean priority.

3

Matthew's and Luke's use of Mark: a logistic regression approach

3.1 Introduction

There is no model of a kind similar to the triple-link model that could be used to assess the validity of the widely accepted two-source hypothesis for synoptic relationships (see Figure 1.2) or to compare its goodness of fit with that of the triple-link model, because we have no surviving text of the hypothetical Q source from which transmission probabilities of words could be calculated in the way that we have done for the triple-link model.

Any attempted reconstruction of a Q text, such as that of Robinson et al. (2000), is based essentially on words that are present in the Matthew-Luke double tradition and by its very nature will not include material that may have been present in Q but has not been utilised by both Matthew and Luke. Furthermore, it is generally supposed by proponents of the two-source hypothesis that there is some overlap between Mark and Q.[1] Examples of pericopes which consist partly of triple tradition and partly of Matthew-Luke double tradition material, and which under the two-source hypothesis have some Mark-Q overlap, may be found in Table 6.11. The reconstruction of any Q text where there is such an overlap is especially tentative. So there is no comprehensive, definitive text of Q available to examine Matthew's and Luke's use of it in the way that we can examine their use of Mark.

In order to attempt some investigation of the two-source hypothesis, assuming Markan priority, we shall in this chapter develop models of Matthew's and Luke's verbal agreements with Mark, in particular with a view to exploring whether Matthew and Luke were independent in their use of Mark. In the standard form of the two-source hypothesis it is assumed that Matthew and Luke worked independently of each other, in the sense of not collaborating or, more especially, neither having the other's text available as a source. Although at first glance this might be taken to suggest that they were statistically independent in the words that they retained unchanged from Mark, this is unlikely to be the case, as already noted in the discussion of the assumptions of the triple-link model in Section 2.3. We might expect the criteria that Matthew

[1]See, for example, Streeter (1924), pp. 186-191, or Tuckett (1996), pp. 31-34.

and Luke used to select words from Mark to have some similarities. What they regarded as important to retain precisely word for word might have some common features, as might what they regarded as superfluous or problematical. Furthermore, they might both have been influenced by similar oral traditions or, perhaps, other written sources that affected their use of Mark in similar ways. The result would be that there would be some departures from statistical independence. If in our statistical analysis we were to find that there was no significant evidence to reject the hypothesis of statistical independence of Matthew and Luke in the words retained unchanged from Mark then we could conclude that we had found no evidence to reject the hypothesis that Matthew and Luke worked independently of each other in their use of Mark. But if, as will turn out to be the case, we find significant evidence to reject the hypothesis of statistical independence then it is less straightforward to decide what we might conclude. To put it rather circumspectly, the stronger the statistical dependence turns out to be, the stronger the support for synoptic models that do not assume the independence of Matthew and Luke in their use of Mark. We should be clear then that the term "(in)dependence" is at this point being used in a different sense from "statistical (in)dependence". So we have to be rather careful in the use of the term "(in)dependence", as it can carry different meanings.[2] In this chapter we are concerned primarily with issues of statistical dependence, which does not necessarily imply dependence in the other sense being used here, of one author having another as a source.

The assumption of Markan priority is implicit in the two-source hypothesis, but does not imply it. As alternatives to the two-source hypothesis, we may consider the two cases of the triple-link model (Figure 2.2) that assume Markan priority but dispense with the need for the Q source. Firstly, there is the case A-B-C = Mk-Mt-Lk that corresponds to the Farrer hypothesis (Figure 1.3), according to which Matthew used Mark, but Luke used both Mark and Matthew as sources. Secondly there is the case Mk-Lk-Mt, according to which Luke used Mark, but Matthew used both Mark and Luke as sources. In either of these two cases of the triple-link model, irrespective of whether we use the original or modified form of the model, and as explicitly shown in Sections 2.4.1. and 2.4.2, the words retained unchanged from Mark by Matthew and Luke are not statistically independent, but instead there is a positive association. In either of these two cases, we could expect there to be a much stronger statistical dependence between the words that Matthew and Luke retained from Mark than is the case with the two-source hypothesis. Taking things a little further, we might look specifically for evidence that the text of Matthew influenced Luke's use of Mark or that the text of Luke influenced Matthew's use of Mark.

There are other models too that assume Markan priority, for example, the three-source hypothesis, which is an extension of the two-source hypothesis

[2] Watson (2013), pp. 156, 217, discusses the term "dependent" from a different perspective that emphasises the dependent author as interpreter of his source.

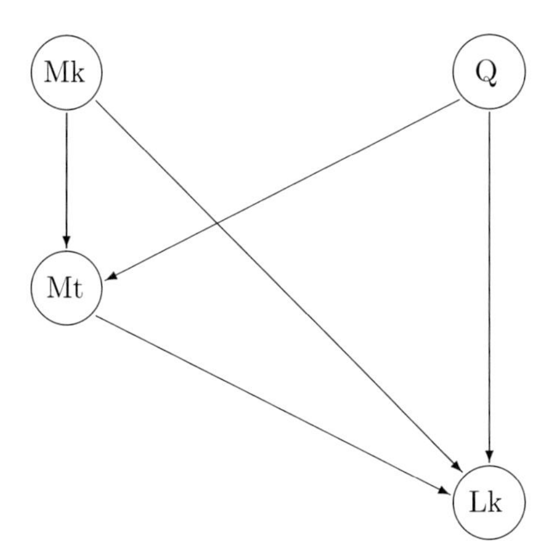

Figure 3.1
The three-source hypothesis

such that Luke also has Matthew as a source. Thus Matthew has Mark and Q as sources and Luke has Mark, Q and Matthew as sources, as illustrated in Figure 3.1. The three-source hypothesis has been advocated by Morgenthaler (1971) and Gundry (1994).[3] This model certainly has greater explanatory power than either the two-source hypothesis or the Farrer hypothesis, but at the expense of a loss of parsimony. Proponents of the two-source hypothesis or the Farrer hypothesis may regard the three-source hypothesis as unnecessarily complicated.

One way of describing the content of this chapter is to say that we focus on a particular component of the two-source hypothesis: the fork model case of the double-link model with Mark as the source, which also featured at the end of Section 2.6. An important development that takes us beyond the material of Chapter 2 is that we construct statistical models of how the sequence of successive words in Mark is utilized by Matthew and Luke, with the result that statistical tests of hypotheses can be carried out.

In this chapter we investigate the nature and extent of the statistical dependence between Matthew's and Luke's use of Mark. In Section 3.2 we describe the data set to be used, which at its heart consists of a bivariate binary time series that represents Matthew's and Luke's verbal agreements and non-agreements with Mark. Before embarking on a more detailed examination of the statistical dependence between Matthew's and Luke's use of Mark, in

[3]See Morgenthaler (1971), pp. 300f., and Gundry (1994), p. 5.

Section 3.3 we attempt to model the way in which Matthew and Luke each individually used the text of Mark. After an exploratory analysis based upon the fitting of variable length Markov chains, logistic regression models are fitted to the binary time series. In Section 3.4 we introduce terms into the logistic regression models that allow for the influence of Luke on Matthew's use of Mark and vice versa. In Section 3.5, the conclusions are summarized regarding the evidence of statistical dependence between Matthew's and Luke's use of Mark.

3.2 The data

As in Chapter 2, the statistical analysis here will be based upon observation of verbal agreements between the synoptic gospels, that is, of common occurrences of the same Greek word in the same context and in the same grammatical form. As discussed in Section 2.7, the results of the analysis of the triple-link model were presented with no formal indication of their statistical significance, but a new feature of the present chapter is that, after constructing a data set that explicitly takes into account the word order in Mark, we shall be able to develop statistical models whose goodness of fit can be compared through their residual deviances and statistical tests of significance.

The data set is a word-by-word transcription of the colour-coded text of Mark in the *Synopticon* of Farmer (1969), described in Section 1.4.2, into a bivariate binary time series that represents Matthew's and Luke's use of Mark. The time series is of length 11078, which is the number of words in the Greek text of Mark that was used by Farmer (but finishing at Mk 16:8 and excluding the longer ending, generally regarded as a later addition to the text). Verbal agreements are coded 1 and non-agreements 0. The first component (x_t) of the bivariate time series is constructed by writing $x_t = 1$ if the word in position t in the text of Mark is present unchanged in Matthew and $x_t = 0$ otherwise. Similarly, the second component (y_t) is constructed by writing 1 if the word is present unchanged in Luke and 0 otherwise. Thus the subscript t of the time series refers to the position of the word in the text of Mark. It should be noted that the value of t does not determine the position of any verbal agreement in Matthew or Luke. Indeed, Matthew and Luke may have changed the relative positions of words from where they are found in Mark.

The data could be regarded as a spatial process in one dimension, but in fact there is a natural direction to the data, the order in which the text was written down by Mark and in which it was read by Matthew and Luke, so that it is more natural to think of the data as a time series, which is what is done in the present approach.

The total numbers of zeros and ones represent overall counts of verbal agreements and non-agreements between Mark and the other synoptic gospels.

Table 3.1
Counts from Farmer's *Synopticon* of verbal agreements with Mark and, in brackets, expected numbers under the hypothesis of statistical independence

		Luke		
		0	1	total
Matthew	0	5243 (4606)	1119 (1756)	6362
	1	2778 (3415)	1938 (1301)	4716
	total	8021	3057	11078

These counts are presented in Table 3.1 in the form of a contingency table, where in each cell, below the observed frequency, we have in brackets, to the nearest integer, the expected frequency for the given row and column totals, under the hypothesis that Matthew and Luke are statistically independent in their verbal agreements with Mark.

The counts in Table 3.1 for the data from Farmer's *Synopticon* differ somewhat from the counts from Honoré's data in Table 2.4 and Table 2.10 and from the counts for the Tyson and Longstaff data in Table 2.5. This occurs for the reasons listed in Section 1.4.2, but the overall picture is much the same for each of the three sets of data. By inspection of the row and column totals in Table 3.1 we see that Matthew follows Mark more closely than does Luke, retaining 4716/11078 = 43% of Mark's words unchanged, whereas Luke retains 3057/11078 = 28% of Mark's words unchanged. These proportions, 0.43 and 0.28, are also the mean values of the series (x_t) and (y_t), respectively.

The observed frequencies along the diagonal of Table 3.1 exceed the expected frequencies, which shows that Matthew and Luke make the same decision on whether or not to retain unchanged a word in Mark more often than would be expected under the hypothesis of statistical independence. As discussed in Section 3.1, we might expect some such positive association under the two-source hypothesis, but we would certainly expect it under the Mk-Mt-Lk or Mk-Lk-Mt cases of the triple-link model. As explained in Section 2.6 with regard to Table 2.10, it is inappropriate to carry out a simple chi-square test, but the data in Table 3.1 do at least suggest that there may be serious evidence that Matthew and Luke are not statistically independent in their verbal agreements with Mark, and our time series analysis will confirm this.[4]

One way in which statistical dependence between Matthew's and Luke's

[4] If at this stage we formally carry out a simple chi-square test with 1 degree of freedom, a procedure that as noted above is invalid, the p-value of the test statistic is of the order of 10^{-164}. This value is so minute as to lead us to expect that, when we carry out a

use of Mark might have arisen, even if they were working independently of each other, is if they used similar criteria in deciding what types of text it was important to retain unchanged. Morgenthaler (1971) in his major statistical analysis of the texts of the gospels distinguished between several types of text. Tyson and Longstaff (1978) too classified sections of text as to whether they were narrative material or words of Jesus or John the Baptist, i.e., material that is often referred to as "sayings".

From the Greek text of the Gospel of Mark it is easy to specify precisely which words make up the direct speech of Jesus. There is also a short piece of the direct speech of John the Baptist and two short pieces of direct speech representing the divine voice from heaven. We have constructed another binary time series (z_t) by writing $z_t = 1$ if the word in position t in the text of Mark is part of the direct speech of Jesus or John or the divine voice and $z_t = 0$ if it is not part of such direct speech. New Testament scholars generally agree that the writers of the gospels and those who transmitted the tradition orally through public performance would have had a greater tendency to reproduce precisely word for word the sayings of Jesus or John but would have felt more at liberty to vary the narrative and editorial material and the speech of other participants in the narrative. This view is substantiated, for example, by the statistics tabulated on p. 71 of Carlston and Norlin (1971). Hence it seems appropriate to introduce z_t as a covariate into our models to investigate the extent to which it helps to explain the variation in the series (x_t) and (y_t) and the dependence between them. For illustration, a section of the series (x_t) and (y_t) together with the covariate series (z_t) is shown in Table 3.2.

As discussed in Section 1.3, the texts of the gospels may be partitioned into sections, known as pericopes. Two standard specifications of the pericopes are provided in their synopses by Huck (1949) and Aland (1996), respectively. We shall make use of the former, which is geared specifically to comparison of the three synoptic gospels and which partitions the Gospel of Mark into 103 pericopes that range in length from 15 to 374 words.[5] To take into account that there may be variation in the way that Matthew and Luke handle the different Markan pericopes, we shall later introduce a factor for pericope into our models.

For the present, to illustrate in outline the way in which the series (x_t) and (y_t) vary over the length of Mark's gospel, in Figure 3.2 we provide a plot of the mean values of x_t (the solid line) and y_t (the dashed line) by pericope, where for the purposes of this plot the pericopes have been numbered 1 to 103 in the order in which they appear in Mark's gospel. These means are, equivalently, the proportions of Mark's words that are retained unchanged by Matthew and Luke, respectively. As may be seen from the plots, there is a great deal of variation in these means among the pericopes. As observed in

---

more sophisticated and appropriate analysis, we might still obtain significant evidence that Matthew and Luke are not statistically independent in their verbal agreements with Mark.

[5] For ease of reference, it may be noted that the English language synopsis of Throckmorton (1992) adopts the Huck specification of pericopes.

Table 3.2
A section (Mk 9:36-37) of the series x_t, y_t, z_t

chapter	verse	word	t	x_t	y_t	z_t
9	36	1	5982	1	0	0
9	36	2	5983	0	0	0
9	36	3	5984	1	1	0
9	36	4	5985	1	1	0
9	36	5	5986	1	1	0
9	36	6	5987	1	0	0
9	36	7	5988	1	0	0
9	36	8	5989	1	0	0
9	36	9	5990	1	1	0
9	36	10	5991	0	0	0
9	36	11	5992	0	0	0
9	36	12	5993	1	1	0
9	36	13	5994	0	1	0
9	37	1	5995	1	1	1
9	37	2	5996	0	0	1
9	37	3	5997	1	0	1
9	37	4	5998	0	0	1
9	37	5	5999	0	0	1
9	37	6	6000	0	0	1
9	37	7	6001	1	1	1
9	37	8	6002	1	1	1
9	37	9	6003	1	1	1
9	37	10	6004	1	1	1
9	37	11	6005	1	1	1
9	37	12	6006	1	1	1
9	37	13	6007	1	1	1
9	37	14	6008	0	1	1
9	37	15	6009	0	1	1
9	37	16	6010	0	1	1
9	37	17	6011	0	1	1
9	37	18	6012	0	0	1
9	37	19	6013	0	0	1
9	37	20	6014	0	0	1
9	37	21	6015	1	1	1
9	37	22	6016	0	0	1
9	37	23	6017	1	1	1
9	37	24	6018	1	1	1
9	37	25	6019	1	1	1

the comments on Table 3.1, the overall mean for (x_t) is 0.43 and for (y_t) is 0.28, and this is reflected in the plot of Figure 3.2, where the solid line tends to lie above the dashed line.

3.3 Models for the univariate series

Before considering the relationship between Matthew's and Luke's use of Mark, as a preliminary, we consider in this section the modelling of the binary series (x_t) for Matthew's verbal agreements with Mark and of the binary series (y_t) for Luke's verbal agreements with Mark, when the series are considered individually, without reference to each other.

When either of the series (x_t) and (y_t) is examined, it becomes apparent that at any point t the probability of a 1 occurring depends on the previous history of the series. A previous run of 0s makes it less likely that there will be 1 in the current position, but a previous run of 1s will make it more likely that there will be a 1 in the current position. In other words, there is some clustering of 1s and of 0s.

One approach to modelling categorical, and in particular binary, time series is by using variable length Markov chains (VLMCs). This method is described, for example, by Mächler and Bühlmann (2004), who also introduce the R package `VLMC` that provides an algorithm for fitting VLMCs. In a VLMC model the order of the Markov chain that is used at any point depends on the history of the process, i.e., the transition probabilities are determined by looking back at a variable number of lagged values of the series. The numbers of lags used in a particular fitted model will depend upon the tuning parameters chosen for the VLMC algorithm.

When the results of applying the VLMC algorithm in R to the series (x_t) and (y_t) were examined, no particularly illuminating models emerged nor was there any clear-cut indication of the number of lags that should be used. What did emerge, however, was that the transition probabilities generated by the models suggested by the VLMC algorithm were based upon the number of 0s since the last occurrence of 1 and, to a lesser extent, the number of 1s since the last occurrence of a 0.

Table 3.3 shows a VLMC model fitted to the series (x_t). In this case the overall order of the fitted Markov chain is 8, the maximum number of lagged values of the series used in the fitted model. The term *context* here refers to the relevant history $x_{t-1}, x_{t-2}, x_{t-3}, \ldots$ of the process at any point t, and the estimated probability, given any particular context, is simply the relative frequency in the observed run of the series of the occurrence of $x_t = 1$ over all occurrences of the given context.

So, viewing the VLMC algorithm as an exploratory technique, what was suggested was that useful predictors of the next value in the series might be

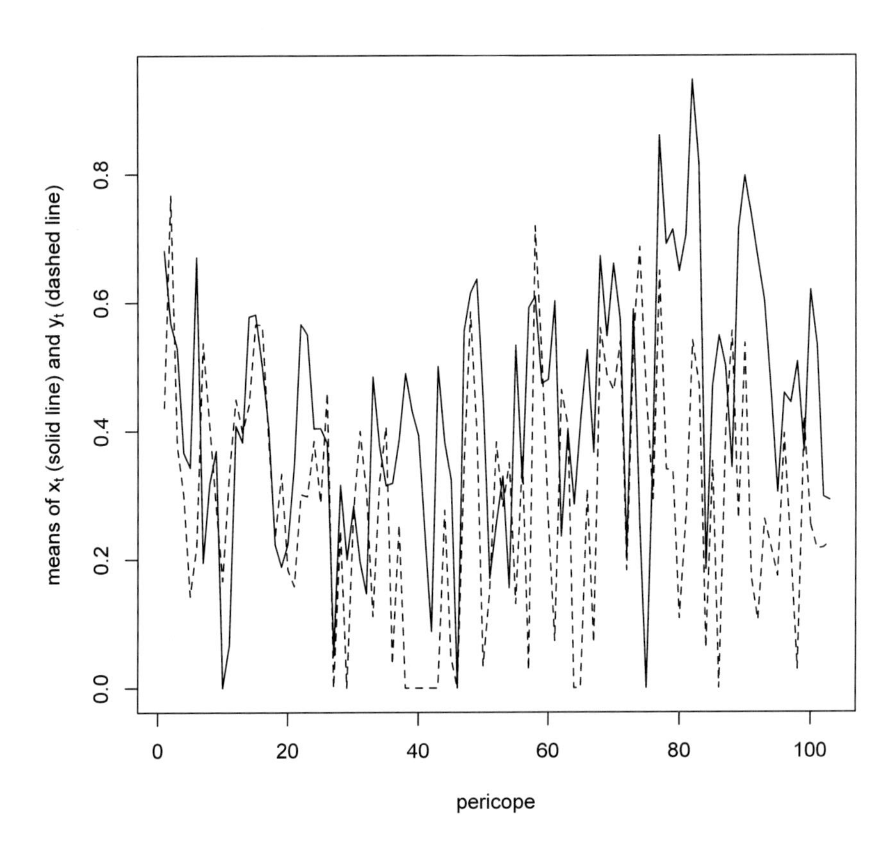

Figure 3.2
Plot of the mean values of x_t and y_t by pericope in Mark

Table 3.3
A fitted VLMC model for the series (x_t)

context $x_{t-1}, x_{t-2}, x_{t-3}, \ldots$	estimated $\Pr(X_t = 1)$
00000000	$172/2221 = 0.077$
00000001	$40/213 = 0.188$
0000001	$62/275 = 0.225$
000001	$67/342 = 0.196$
00001	$126/468 = 0.269$
0001	$145/613 = 0.237$
001	$254/867 = 0.293$
01	$489/1356 = 0.361$
10	$871/1356 = 0.642$
110	$608/871 = 0.698$
1110	$423/608 = 0.696$
1111	$1458/1881 = 0.775$

the current run lengths of 0s and 1s, or some function of them, and that these would provide a compact way of representing the effect of the history of the process upon the probability distribution of the next value, perhaps to a large extent replacing what might otherwise be a complicated function of several lagged values and their interactions.

For the main part of our analysis in this chapter, we use generalized linear modelling, which in a time series setting is presented in Kedem and Fokianos (2002), where the use of the standard methods of generalized linear modelling, as provided by McCullagh and Nelder (1989), is justified for the analysis of time series through a partial likelihood approach. Kedem and Fokianos in their Chapter 2 deal specifically with the case of binary time series, including the use of logistic regression.

Assuming that the series has been observed up to the $(t-1)$th position, or, equivalently, in the language of time series, assuming that the series has been observed up to time $t-1$, let π_t denote the probability that there is a 1 in position t. More formally, for the series (x_t),

$$\pi_t = \Pr(X_t = 1|\mathcal{F}_{t-1}),$$

where the upper case X_t represents the binary random variable at time t and $\mathcal{F}_{t-1}$ the history of the series up to time $t-1$.

Let N_t^0 denote the current run of 0s at time t and N_t^1 denote the current run of 1s, where one or other of N_t^0 and N_t^1 will always be zero. From the exploratory analysis using VLMCs, it was found that N_{t-1}^0 and N_{t-1}^1 might be especially important predictor variables for π_t. In fact, some further investigation showed that better predictor variables, as judged by comparison of the residual deviances of the fitted models, were given by taking logarithms and using R^0_{t-1} and R^1_{t-1}, where

$$R^0_t = \ln(1 + N^0_t)$$

and

$$R^1_t = \ln(1 + N^1_t) \ .$$

It was also anticipated that, in addition, the recent history of the process might be particularly influential so that, to supplement the information in the variables R^0_{t-1} and R^1_{t-1}, a small number of the lagged variables X_{t-1}, X_{t-2}, ... might also be used as predictors, and possibly their interactions. Using other link functions in the generalized linear model appeared to do no better than using the canonical logit link,[6] so the model adopted was a logistic regression of the form

$$\ln\left(\frac{\pi_t}{1-\pi_t}\right) = \alpha + \beta_0 R^0_{t-1} + \beta_1 R^1_{t-1} + \gamma_1 X_{t-1} + \gamma_2 X_{t-2}, \qquad (3.1)$$

but envisaging the possibility that not all the terms would be needed or that some further lagged terms and interactions might be added.

Similarly, if θ_t is the probability that there is a 1 in position t for the series (y_t), i.e.,

$$\theta_t = \Pr(Y_t = 1 | \mathcal{G}_{t-1}),$$

where $\mathcal{G}_{t-1}$ represents the history of the series (y_t) up to time $t-1$, the model adopted was of the form

$$\ln\left(\frac{\theta_t}{1-\theta_t}\right) = \alpha + \beta_0 S^0_{t-1} + \beta_1 S^1_{t-1} + \gamma_1 Y_{t-1} + \gamma_2 Y_{t-2}, \qquad (3.2)$$

where S^0_t and S^1_t are the logarithms of the run lengths for the series (y_t), defined in exactly the same way as R^0_t and R^1_t for the series (x_t).

As a check on the appropriateness of regressing the logits on the logarithms of run lengths for the series (x_t), simple estimates $\hat{\pi}_t$ of $\Pr(X_t = 1)$ were calculated conditional upon the values of the run lengths N^0_{t-1} and N^1_{t-1}, using as estimates the values of relative frequencies, just as in Table 3.3 for the VLMC model. In Figure 3.3 the logits of $\hat{\pi}_t$ have been plotted against values of R^0_{t-1} and R^1_{t-1}. Using a similar calculation for the series (y_t), the logits of $\hat{\theta}_t$ have been plotted against values of S^0_{t-1} and S^1_{t-1}. The plots appear

[6]The ratio of π_t to $1-\pi_t$ is the *odds* that $X_t = 1$ occurs, so that the *logit* (or *logistic*) transformation

$$\ln\left(\frac{\pi_t}{1-\pi_t}\right)$$

is the natural logarithm of the odds. The probability π_t is restricted to take values in the interval (0,1), but the logit can take any value, positive or negative. Like the probability π_t itself, the logit may be regarded as a measure of how likely it is that $X_t = 1$ occurs, only measured on a different scale.

to be reasonably linear except for the zero values of the predictor variable, but these are special values because, for example, when one of R^0_{t-1} and R^1_{t-1} is zero and absent from the regression then the other is non-zero and contributes to the regression. Furthermore, the possible presence of the predictor variables $X_{t-1}, X_{t-2}, \ldots$ and interaction terms may effect a further adjustment to the regression if this turns out to be necessary.

Models of the form of Equation (3.1) and Equation (3.2) were fitted to the series (x_t) and (y_t), respectively, using the `glm` and related functions in R. In particular, the `step` function was used for stepwise selection of predictor variables starting from the null model with no predictor variables. In practice the method led to forward selection of variables.[7] The `step` function is based on the use of the Akaike information criterion (AIC), but it also outputs the values of residual deviance at each step, which enables tests of significance to be carried out, based on the asymptotic likelihood ratio test.

When models of the form of Equation (3.1) were fitted to the series (x_t) that represents Matthew's use of Mark, it was found that the best single predictor to use was R^0_{t-1}. A highly significant improvement in fit was obtained by also including R^1_{t-1} as a predictor. A further significant improvement was obtained by including X_{t-1}, but then no significant improvement was obtained by introducing further lagged variables. It should be recalled that for binary data it is not appropriate to use the residual deviance as an absolute measure of the goodness of fit of the model. See Kedem and Fokianos (2002), p. 66, and McCullagh and Nelder (1989), pp. 121-122. So the question of how well the model fits the data is left somewhat open, although it is appropriate to look at changes in residual deviance when assessing the significance of introducing additional terms into the model. The estimated regression coefficients for this model (Mt1) are given in the first column of numbers in Table 3.4. Thus the corresponding fitted model equation is

$$\ln\left(\frac{\pi_t}{1-\pi_t}\right) = 0.017 - 0.806R^0_{t-1} + 0.441R^1_{t-1} + 0.285X_{t-1}.$$

We may note that the signs of the coefficients are as expected. The probability of observing a 1 in the current position increases with the length of a previous run of 1s and decreases with the length of a previous run of 0s.

A similar fitted model (Lk1) of the form of Equation (3.2) emerges for the series (y_t) that represents Luke's use of Mark. Its estimated regression coefficients are given in the first column of numbers in Table 3.5. In what follows, as further variables are introduced into the regression equations, each column in Table 3.4 and Table 3.5 will give the estimated regression coefficients for the model chosen as a result of a stepwise procedure for the current set of candidate predictor variables, but the corresponding fitted model equation

[7] Stepwise methods fit a sequence of models, at each stage adding or removing a predictor variable to obtain a better model according to a specified criterion, until the procedure converges to a final suggested model. Forward selection is where predictor variables are added one at a time, but none removed.

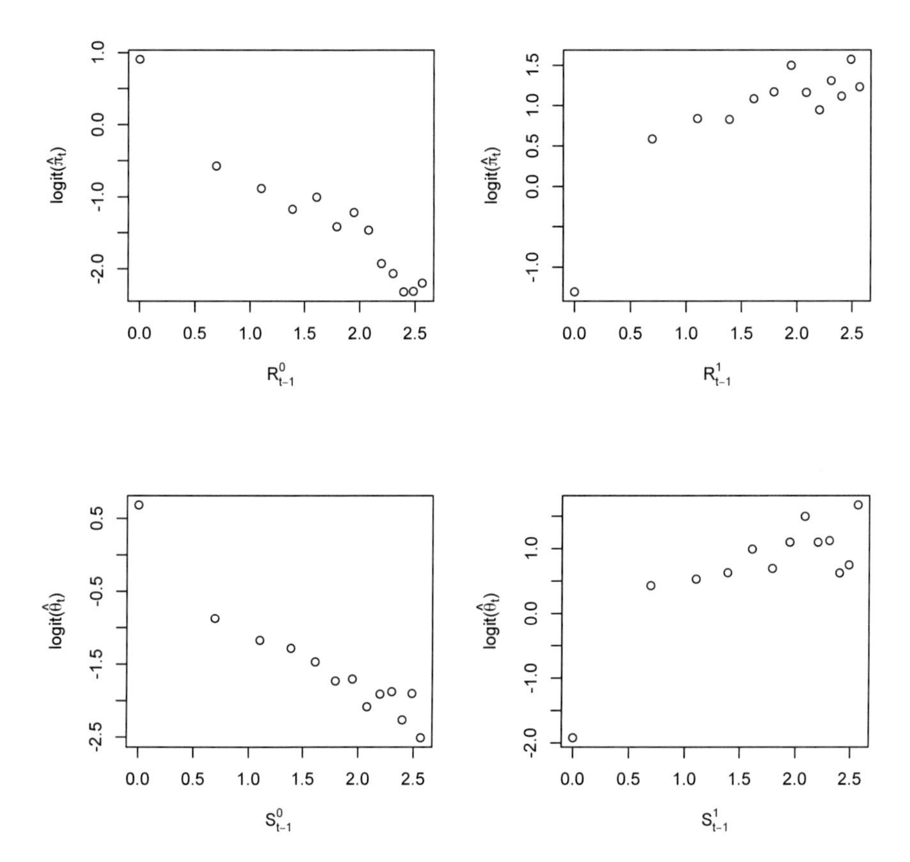

Figure 3.3
Plots of logits against logarithms of run lengths

Table 3.4
Estimated regression coefficients for the series (x_t)

Model	Mt1	Mt2	Mt3	Mt4	Mt5	Mt6
constant	0.017	0.005	-0.160	-0.300	-0.393	-0.477
R^0_{t-1}	-0.806	-0.867	-0.864	-0.889	-0.885	-0.823
R^1_{t-1}	0.441	0.511	0.522	0.573	0.551	0.512
X_{t-1}	0.285	–	–	–	–	–
Z_t	–	0.399	–	–	0.339	0.399
Y_t	–	–	1.090	1.441	1.419	1.468
S^0_{t-1}	–	–	–	0.072	0.067	0.062
S^1_{t-1}	–	–	–	-0.301	-0.323	-0.289
pericope factor	–	–	–	–	–	✓
residual deviance	11675	11612	11229	11117	11069	10993
residual d.f.	11072	11072	11072	11070	11069	(11067)

will not be written out. Standard errors for the regression coefficients are not included in Tables 3.4 and 3.5 but are shown in Tables 3.6 and 3.7 for the final selected models.

We next consider introducing the covariate series (z_t) for direct speech and using Z_t as an additional predictor variable for the series (x_t) and (y_t). When modelling the series (x_t), we find again that the best pair of predictors to use is R^0_{t-1} and R^1_{t-1}, but the next variable that provides the greatest improvement in fit is Z_t. The estimated regression coefficients for the resulting model (Mt2) are given in Table 3.4. Further significant but small improvements in fit are given by introducing the interaction of Z_t with R^1_{t-1} and then X_{t-1} into the model. However, we have chosen to present the simpler model Mt2, that corresponds to terminating the stepwise procedure after three steps, in Table 3.4. As further predictor variables are introduced in Section 3.4, the models become increasingly complex, and for ease of presentation and interpretation it was decided at this stage to keep to a simpler model. In so doing, nothing essential to the argument in Section 3.4 is lost.

Similarly, when modelling the series (y_t), we find again that the best pair of predictors to use is S^0_{t-1} and S^1_{t-1}, but the next variable that provides the greatest improvement in fit is Z_t. The estimated regression coefficients for the resulting model (Lk2) are given in Table 3.5. Further significant but small improvements in fit are given by introducing the interaction of Z_t with S^0_{t-1} and then Y_{t-1} into the model.

We see from the residual deviances that the covariate Z_t for direct speech does give some improvement in fit for the univariate series. As expected, the estimated regression coefficients for Z_t are positive. Clearly, a word that is a part of direct speech is more likely to be retained unchanged by Matthew or Luke than a word that is a part of the narrative.

Table 3.5
Estimated regression coefficients for the series (y_t)

Model	Lk1	Lk2	Lk3	Lk4	Lk5	Lk6
constant	-0.257	-0.284	-0.697	-0.954	-1.029	-1.147
S^0_{t-1}	-0.802	-0.851	-0.846	-0.870	-0.872	-0.806
S^1_{t-1}	0.510	0.600	0.607	0.677	0.663	0.636
Y_{t-1}	0.284	–	–	–	–	–
Z_t	–	0.380	–	–	0.280	0.362
X_t	–	–	1.133	1.452	1.430	1.416
R^0_{t-1}	–	–	–	0.199	0.201	0.155
R^1_{t-1}	–	–	–	-0.149	-0.174	-0.164
pericope factor	–	–	–	–	–	✓
residual deviance	9201	9156	8745	8643	8618	8566
residual d.f.	11072	11072	11072	11070	11069	(11067)

The remaining models, Mt3, ..., Mt6 and Lk3, ..., Lk6, in Table 3.4 and Table 3.5, respectively, include terms that model the statistical dependence between the series (x_t) and (y_t) and will be discussed in Section 3.4.

3.4 Models for the bivariate series

We now consider the two series (x_t) and (y_t) as a bivariate time series (x_t, y_t). In so doing we are considering in conjunction Matthew's and Luke's use of Mark and their possible use of each other.

One type of approach that might naturally be considered here is the modelling of the joint distribution of X_t and Y_t in terms of the histories of the processes up to time $t-1$. We could consider a bivariate logistic model as done in Sections 6.5.6 and 6.5.7 of McCullagh and Nelder (1989), and as put in the more general setting of vector generalized additive models by Yee and Wild (1996) and implemented in the R package `VGAM`. Such an approach is taken specifically for certain types of bivariate binary time series by Mosconi and Seri (2006), though using a probit rather than a logit link function.

However, the specific setting here, where we have in mind the possibility that Matthew is using Luke or Luke is using Matthew as a source, suggests that it is more natural to model the distributions of X_t and Y_t separately: X_t not only in terms of the history $\mathcal{F}_{t-1}$ of the series (x_t) itself up to time $t-1$ but also in terms of the history $\mathcal{G}_t$ of the series (y_t) up to time t, including, importantly, the current value Y_t; and, similarly, Y_t not only in terms of the history $\mathcal{G}_{t-1}$ of the series (y_t) itself up to time $t-1$ but also in terms of the history $\mathcal{F}_t$ of the series (x_t) up to time t, including the current value X_t. Furthermore, it may be illuminating to consider our analysis in relation to the concept of causality in the sense that it is discussed in the econometric literature, where causality is expressed in terms of prediction. In particular, using the terminology of Granger (1969), there is *instantaneous causality* of (y_t) acting on (x_t) if the current value of X_t is better predicted when the current value Y_t is included as a predictor variable than when it is not. It should be noted, though, that even if it is found that there is causality in this specific sense, this will not establish that Luke is a source for Matthew, although it may lend support to such a hypothesis. Similarly, if Y_t is better predicted when the current value X_t is included as a predictor variable, this will not establish that Matthew is a source for Luke, although, again, it may lend support to such a hypothesis.

Adopting this approach, when modelling the series (x_t) that represents Matthew's use of Mark, we consider as predictor variables for X_t not only the variables used in Section 3.3 that are functions of $\mathcal{F}_{t-1}$ but also the corresponding variables that are functions of $\mathcal{G}_{t-1}$ and, additionally, the current value Y_t. For the present, we do not use the covariate Z_t. When variables were entered stepwise into the model equation, it was found, as in Section 3.3, that the best single predictor to use was R^0_{t-1}, but the next best variable to enter was Y_t, and only at the third step did the variable R^1_{t-1} enter into the equation. All three of these variables provided a highly significant contribution to the fit. The estimated regression coefficients for the resulting model, Mt3, are given in Table 3.4. Comparison of the residual deviances shows that model Mt3 gives a substantial improvement in fit over the model Mt1, and like model Mt1 it has a simple natural interpretation: the probability of a word in Mark being retained unchanged by Matthew decreases as the length of a previous run of non-retentions increases and increases as the length of a previous run of retentions increases, and also increases if the word is retained unchanged by Luke. Here there is indeed instantaneous causality of (y_t) acting on (x_t), a statistical dependence between X_t and Y_t in the presence of the other predictor variables being used. Further highly significant improvements in fit are found by bringing in additional variables from $\mathcal{G}_{t-1}$ to obtain a model Mt4, whose estimated regression coefficients are given in Table 3.4, whereas bringing in the variable X_{t-1} gives only a relatively small improvement in fit. However, the signs of the estimated regression coefficients for the additional terms S^0_{t-1} and S^1_{t-1} in the model Mt4 are rather puzzling in that we might have expected the probability of finding a retention in Matthew to decrease with the length of a previous run of non-retentions in Luke and to increase with the length of a previous run of retentions in Luke.

A very similar scenario emerged when models were fitted to the series (y_t) that represents Luke's use of Mark, considering for Y_t the same predictor variables as before that are functions of $\mathcal{G}_{t-1}$ and $\mathcal{F}_{t-1}$ and, additionally, the current value X_t. When variables were entered stepwise into the model equation, it was found, as in Section 3.3, that the best single predictor to use was S^0_{t-1}, but the next best variable to enter was X_t, and only at the third step did the variable S^1_{t-1} enter into the equation. All three of these variables provided a highly significant contribution to the fit. The estimated regression coefficients for the resulting model, Lk3, are given in Table 3.5. Comparison of the residual deviances shows that model Lk3 gives a substantial improvement in fit over the model Lk1. So again we have instantaneous causality, now of (x_t) acting on (y_t). Further highly significant improvements in fit are found by bringing in additional variables from $\mathcal{F}_{t-1}$ to obtain a model Lk4, whose estimated regression coefficients are given in Table 3.5, whereas bringing in the variable Y_{t-1} gives only a relatively small improvement in fit. As for the model Mt4, the signs of the estimated regression coefficients for the additional terms R^0_{t-1} and R^1_{t-1} in the model Lk4 are the opposite of what might have been expected.

An important question concerns the extent to which the introduction of the covariate Z_t for direct speech into the models Mt4 and Lk4 will be able to account for the statistical dependence between whether a word is retained unchanged by Matthew and whether it is retained unchanged by Luke. We now consider the series (x_t) and (y_t) using the same predictor variables as in the models Mt4 and Lk4 but with the addition of the variable Z_t. When modelling the series (x_t) using a stepwise approach, we find again that the predictor variables for X_t enter into the model equation in the order R^0_{t-1}, Y_t, R^1_{t-1}, with Z_t entering only at step 5. The model Mt4 with the addition of Z_t as a predictor variable gives the model Mt5 with estimated regression coefficients as given in Table 3.4. The introduction of the variable Z_t does give a significant improvement in fit but has very little impact on the conclusion that the probability that a word is retained unchanged by Matthew is strongly dependent upon whether it is retained unchanged by Luke. Similarly, when modelling the series (y_t) using a stepwise approach, we find again that the predictor variables for Y_t enter into the model equation in the order S^0_{t-1}, X_t, S^1_{t-1}, with Z_t entering only at step 5. The model Lk4 with the addition of Z_t as a predictor variable gives the Model Lk5 with estimated regression coefficients as given in Table 3.5. Just as when modelling the series for Matthew, so when modelling the series for Luke, we find that the introduction of the variable Z_t has very little impact on the conclusion that the probability that a word is retained unchanged by Luke is strongly dependent upon whether it is retained unchanged by Matthew.

In a further attempt to find a way of accounting for the dependence between the series (x_t) and (y_t), in addition to the predictor variables used in the models Mt5 and Lk5, we introduce a normally distributed random effect $B_{H(t)}$ for pericope, where $H(t)$ denotes the pericope according to the specification of Huck (1949) to which the word in position t belongs. The factor for pericope has 103 levels, and we may envisage the pericopes in Mark as being a selection of units of material from a much larger body of material that was available in the oral tradition. So it seems not inappropriate to treat the pericope as a random factor. In addition, because we are especially interested in

Table 3.6
Estimated regression coefficients and standard errors for the model Mt6 for (x_t)

Variable	estimated coefficient	standard error	odds ratio
constant	-0.477	0.087	
R^0_{t-1}	-0.823	0.037	
R^1_{t-1}	0.512	0.043	
Y_t	1.468	0.075	4.342
S^0_{t-1}	0.062	0.023	
S^1_{t-1}	-0.289	0.049	
Z_t	0.399	0.059	1.491

the dependence between (x_t) and (y_t) and how it might vary from pericope to pericope, we also introduce a normally distributed random interaction effect between Y_t and the pericope $H(t)$ into the model for (x_t) and, similarly, a normally distributed random interaction effect between X_t and $H(t)$ into the model for (y_t). Hence we are now dealing with generalized linear mixed models, which we fit using the `lmer` function in the `lme4` package in R, a function which uses a method of penalized least squares for fitting the model. The resulting model Mt6 for the series (x_t) has an estimated standard deviation of 0.380 for the main random effect, an estimated standard deviation of 0.317 for the interaction random effect, and estimated regression coefficients as given in Table 3.6 together with their standard errors. The corresponding odds ratios for the binary predictor variables Y_t and Z_t are also given in Table 3.6. The model Lk6 for the series (y_t) has an estimated standard deviation of 0.367 for the main random effect, an estimated standard deviation of 0.459 for the interaction random effect, and estimated regression coefficients as given in Table 3.7 together with standard errors and odds ratios. It has been noted, for example by Hartzel et al. (2001), p. 91, that the kind of algorithm used in the `lmer` function may lead to serious bias in the estimates of the regression parameters in logistic models. In the present case, however, given the above caveat, the coefficient 1.468 for Y_t in the model Mt6 and the coefficient 1.416 for X_t in the model Lk6 are both overwhelmingly significant as may be seen by comparing the estimated coefficients with their standard errors.

In both these models, the introduction of the random pericope effect significantly improved the fit of the model, and the further introduction of the random interaction also significantly improved the fit, although to a lesser extent. It should be noted that the usual asymptotic likelihood ratio test for fixed effects models, based on the chi-square distribution, is not applicable to tests of variance components for mixed models, as discussed for example in Stram and Lee (1994) and Visscher (2006). The appropriate distribution of the

Table 3.7
Estimated regression coefficients and standard errors for the model Lk6 for (y_t)

Variable	estimated coefficient	standard error	odds ratio
constant	-1.147	0.099	
S^0_{t-1}	-0.806	0.035	
S^1_{t-1}	0.636	0.052	
X_t	1.416	0.084	4.119
R^0_{t-1}	0.155	0.037	
R^1_{t-1}	-0.164	0.047	
Z_t	0.362	0.068	1.437

test statistic is instead a mixture of chi-square distributions. The bracketed degrees of freedom in the final column of Table 3.4 and Table 3.5, calculated by simply considering the number of fitted parameters, whether for fixed effects or variance components, should be considered as rough guides that suggest chi-square tests that are more conservative than the ones based on mixtures of chi-square distributions.[8] In any case, the results here for the pericope and interaction effects are significant.

For both series, (x_t) and (y_t), the addition of the random effects significantly improved the fit of the model, but in neither case did it have any impact on the conclusion that there is a highly significant statistical dependence between whether a word in Mark is retained unchanged by Matthew and whether it is retained unchanged by Luke.

A very useful by-product of the analysis of the models Mt6 and Lk6 is that we can examine the interactions with the pericope factor of the predictors Y_t and X_t, respectively. Figure 3.4 gives a scattergram of the predicted interactions for the individual pericopes. Those pericopes for which these interactions are largest in a positive direction are the ones where the dependence between the series (x_t) and (y_t) appears to be the strongest. It is these pericopes that are suggested by our analysis as the ones which in the first instance might appear to offer the most serious challenge for defenders of the two-source hypothesis and for which a detailed analysis of the text might be particularly relevant with regard to agreements between Matthew and Luke in what to retain and what to omit or alter. The cluster of five pericopes in the top right-hand corner of Figure 3.4 are presented in Table 3.8 with the pericope numbers as specified in Huck (1949) and Throckmorton (1992).

[8]See Visscher (2006), p. 493.

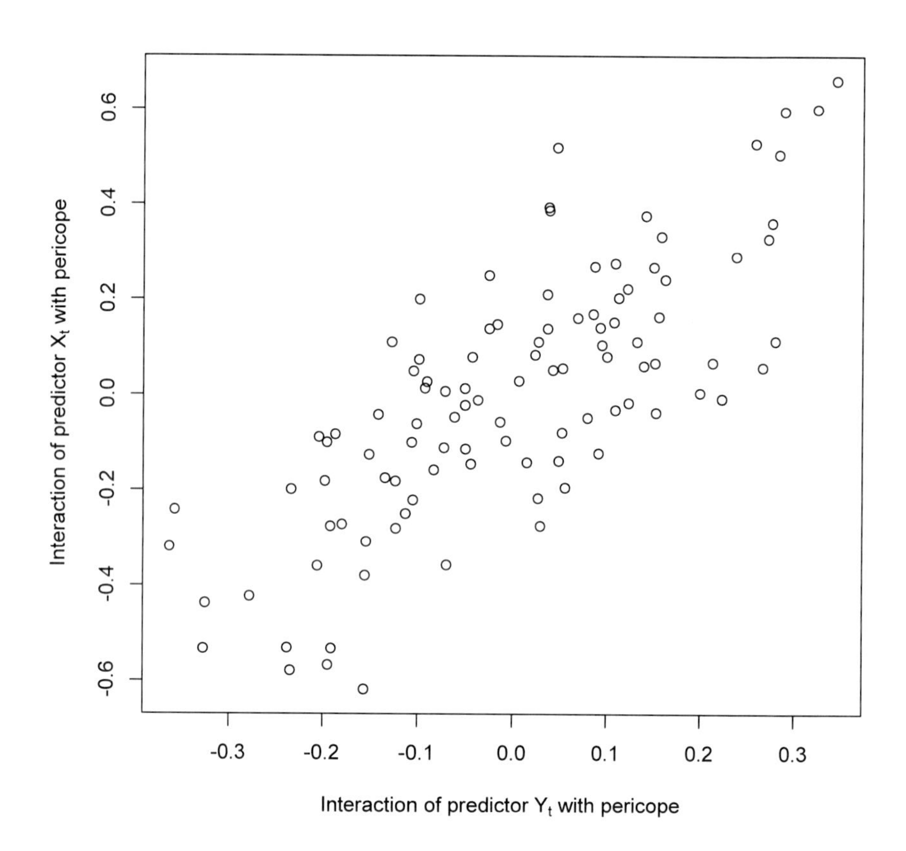

Figure 3.4
Scatter plot of interactions of the predictors X_t, Y_t with the pericope effect

Table 3.8
Pericopes with large positive interactions

Huck number	Pericope description	Passage in Mark	Interactions: with Y_t in Mt6	with X_t in Lk6
45	The Healing of a Leper	1:40-45	0.343	0.661
85	The Pharisees' Accusation	3:20-22	0.283	0.505
129	Dispute about Greatness	9:33-37	0.289	0.596
208	The Great Commandment	12:28-34	0.323	0.600
209	About David's Son	12:35-37a	0.257	0.528

3.5 Conclusions

In summary, on examining Table 3.4 for models of the form of Equation (3.1), we see that in the models from Mt3 onwards, where Y_t is included as a predictor variable, as additional predictor variables or the random factor for pericope are introduced, there is at each step a significant improvement in fit as expressed by a significant decrease in the residual deviance, but the effect of Y_t on predictions of X_t is either increased or only slightly diminished. Similarly, on examining Table 3.5 for models of the form of Equation (3.2), in the models from Lk3 onwards, where X_t is included as a predictor variable, as additional predictor variables or the random factor for pericope are introduced, there is at each step a significant decrease in the residual deviance, but the effect of X_t on predictions of Y_t is either increased or only slightly diminished. In Table 3.4 and Table 3.5, only a few of the best fitting models have been presented, but in these and all other cases examined Y_t and X_t have highly significant positive effects as predictors for each other, whatever further predictor variables are included.

So it appears that there is a very strong statistical dependence between the series (x_t) and (y_t) even when allowance is made for a number of other covariates. In order to understand in more depth the nature of the dependence it is necessary to go down to the level of studying the Greek text in detail and discussing what the reasons might be for why Matthew and Luke tend to agree more often than would be expected by chance on what words of Mark to retain and what to omit or alter. This is the task of New Testament scholars, and we shall be looking at some examples in Chapter 7. At this stage, the pericopes in Table 3.8 are suggested for more detailed investigation, as they appear likely to provide particularly strong evidence of dependence between Matthew's and Luke's use of Mark. Supporters of the two-source hypothesis tend to argue that the apparent dependence is due to similarities in the editorial strategies of Matthew and Luke, which are amenable to rational explanation, or to the influence of similar oral traditions that were available to both of them. Supporters of a triple-link model with Markan priority argue that it is much more natural to explain the agreements by assuming that Luke also had Matthew as a source or vice versa.

We have found in the models fitted in Section 3.4 that there is a strong statistical dependence, a positive association, between whether a word in Mark is retained unchanged by Matthew and whether it is retained unchanged by Luke: if a word has been kept unchanged by Matthew then this makes it more likely that it was kept unchanged by Luke, and vice versa. Such a positive association is natural for theories that assume either that Luke had Matthew's gospel as a source or that Matthew had Luke's gospel as a source, but it is more problematic for the two-source hypothesis according to which Matthew and Luke used Mark independently of each other. Our fitted models for Matthew's use of Mark and for Luke's use of Mark are very similar in form, and our results are inconclusive as to whether it is more likely that Matthew had Luke as a source or that Luke had Matthew as a source.

As discussed briefly above and in Sections 3.1 and 3.2, a hypothesis that Matthew and Luke worked independently of each other could still lead to statistical dependence in the choice of words that they each retained unchanged from Mark. However, the introduction of the covariate for direct speech and of the factor for pericope in Section 3.4 as the most immediately obvious way of accounting for some of the statistical dependence had little effect. A statistical analysis is no substitute for the kinds of detailed analysis of the texts that are carried out by New Testament scholars, but it may be helpful in clarifying certain issues and, as in the present case, raising questions that should perhaps be addressed more comprehensively than has previously been the case. How do proponents of the two-source hypothesis account for the apparently strong and widespread statistical dependence between the texts of Matthew and Luke in their verbal agreements with Mark? From a statistical viewpoint, there is the question of what other ways might be found of modelling the patterns of word retention that would better illuminate or explain the statistical dependence. As one attempt to address this question, in Chapter 4 and the following chapters we explore the use of hidden Markov models.

4

Hidden Markov models for binary time series

4.1 Introduction and motivation

For the present still working with the assumption of Markan priority, we now consider an alternative approach to the one adopted in Chapter 3 for modelling the binary time series (x_t) and (y_t) that represent Matthew's and Luke's verbal agreements with Mark, respectively. The underlying idea is that, when Mark was being used as a source by either Matthew or Luke, different parts of Mark's text were handled in different ways. For example, some sections of Mark's text were followed quite closely, almost word for word, but other sections were followed rather loosely. In some sections of text, Mark might have been essentially the only source for Matthew or Luke, whereas in other sections substantial use of additional sources, including oral tradition, might have been made. Furthermore, some sections of Mark's text appear not to have been used at all. For example, the Markan pericope Mk 8:22-26, "The Blind Man of Bethsaida", is omitted by both Matthew and Luke, and, most strikingly of all, there is Luke's great omission, where he has omitted all the Markan pericopes in the section of text from Mk 6:45-8:10.

So we suppose that behind the observed data as represented by our time series there lie different modes of handling the text of Mark, which are in what follows referred to as hidden states — states which are not directly observable but which we may attempt to infer from the observed data.[1] As Matthew and Luke worked through the text of Mark, they would from time to time switch from one hidden state to another, i.e., from one mode of behaviour to another. In the hidden Markov model approach, the word-by-word sequence of underlying hidden states is modelled as a Markov chain. The modelling process involves the choice of what appears to be the most appropriate number of hidden states to use and estimation of the transition matrix of the Markov chain. At the same time, the behaviour of the observed time series in each hidden state is modelled and the model parameters estimated, which, if things work out well, will lead to a meaningful interpretation for each state.

In Chapter 5 we shall return to the problem of Matthew's and Luke's use

[1]In other branches of statistics, the term *latent* is used instead of *hidden* to describe unobserved variables that underlie the observed ones. For an overview see Bartholomew et al. (2011).

of Mark, but in the present chapter we outline the general theory of hidden Markov models for analysing time series, focusing especially on the case of binary time series. To a large extent we shall follow the treatment given by Zucchini and MacDonald (2009). However, they focused more on series of counts that could be modelled using Poisson distributions, but we shall adapt some of their results to deal with the case of binary time series and include some additional features to be used for the present application in Chapters 5 and 6.

4.2 Description of a hidden Markov model (HMM)

Consider a sequence of random variables, $X_1, X_2, \ldots, X_t, \ldots$, where $t = 1, 2, \ldots$ usually represents successive, equally spaced points in time. The random variable X_t represents the observed value of a time series at time t. The X_t could be either continuous or discrete random variables, but we shall consider here the case of discrete random variables. In particular, we shall restrict attention to binary random variables, where each of the X_t takes either the value 0 or 1, and so the observed time series is just a sequence of 0s and 1s. Thus $\{X_t : t \geq 1\}$ represents the observed binary time series.

Let $\{C_t : t \geq 1\}$ denote a time-homogeneous m-state Markov chain, a sequence of hidden states, which are not directly observable but which influence the observed sequence $\{X_t : t \geq 1\}$.

Let $\mathbf{X}^{(t)}$ denote the history of the process $\{X_t : t \geq 1\}$ up to time t, i.e.,

$$\mathbf{X}^{(t)} = (X_1, X_2, \ldots, X_t).$$

(For convenience, in what follows we shall also use the notation $\mathbf{X}^{(0)}$, to be understood as vacuous, representing the absence of any history.) Similarly, let $\mathbf{C}^{(t)}$ denote the history of the Markov chain $\{C_t : t \geq 1\}$ up to time t, i.e.,

$$\mathbf{C}^{(t)} = (C_1, C_2, \ldots, C_t).$$

In the basic form of the HMM it is assumed that

$$\Pr(C_t | \mathbf{X}^{(t-1)}, \mathbf{C}^{(t-1)}) = \Pr(C_t | C_{t-1}) \qquad (t = 2, 3, \ldots), \qquad (4.1)$$

which is just a form of the Markov property for $\{C_t : t \geq 1\}$, and that

$$\Pr(X_t | \mathbf{X}^{(t-1)}, \mathbf{C}^{(t)}) = \Pr(X_t | C_t) \qquad (t = 1, 2, \ldots), \qquad (4.2)$$

which expresses the assumption that, given the current hidden state C_t, the probability distribution of X_t does not depend on the previous histories $\mathbf{X}^{(t-1)}$ and $\mathbf{C}^{(t-1)}$. In fact, we shall go beyond the basic form of the HMM and, as in the models of Chapter 3, allow the probability distribution of X_t to depend on the previous history $\mathbf{X}^{(t-1)}$ and possibly on the history $\mathbf{Z}^{(t)}$ of some exogenous covariate process $\{Z_t : t \geq 1\}$, which might be a vector process. We shall assume that

$$\Pr(C_t|\mathbf{X}^{(t-1)}, \mathbf{Z}^{(t)}, \mathbf{C}^{(t-1)}) = \Pr(C_t|C_{t-1}) \qquad (t = 2, 3, \ldots), \tag{4.3}$$

and that

$$\Pr(X_t|\mathbf{X}^{(t-1)}, \mathbf{Z}^{(t)}, \mathbf{C}^{(t)}) = \Pr(X_t|\mathbf{X}^{(t-1)}, \mathbf{Z}^{(t)}, C_t) \ \ (t = 1, 2, \ldots), \tag{4.4}$$

so that Equations (4.3) and (4.4) replace Equations (4.1) and (4.2) of the basic model, respectively. As indicated in Chapter 8 of Zucchini and MacDonald (2009), such extensions to the basic model are readily accommodated within the theory of HMMs.

To specify this HMM, we shall need to specify the transition matrix of the Markov chain $\{C_t : t \geq 1\}$. Denoting the states by $1, 2, \ldots, m$, let $\mathbf{\Gamma} = (\gamma_{ij})$ be the transition matrix, the $m \times m$ matrix such that

$$\gamma_{ij} = \Pr(C_t = j|C_{t-1} = i) \qquad (i = 1, 2, \ldots m, j = 1, 2, \ldots, m). \tag{4.5}$$

We may also need to specify an initial probability vector $\boldsymbol{\delta} = (\delta_1, \delta_2, \ldots, \delta_m)$, such that

$$\delta_i = \Pr(C_1 = i) \qquad (i = 1, 2, \ldots m).$$

Assuming that the transition matrix is irreducible and ergodic, as in practice it will be, we may well wish to take the Markov chain $\{C_t : t \geq 1\}$ to be stationary, in which case we take $\boldsymbol{\delta}$ to be the stationary distribution. Then $\boldsymbol{\delta}$ is the unique solution of the equation

$$\boldsymbol{\delta}\mathbf{\Gamma} = \boldsymbol{\delta}$$

that satisfies the normalization condition $\sum_{i=1}^{m} \delta_i = 1$, i.e.,

$$\boldsymbol{\delta}\mathbf{1}' = 1,$$

where $\mathbf{1}$ denotes the $1 \times m$ row vector of 1s and $\mathbf{1}'$ its transpose, the $m \times 1$ column vector of 1s. In that case there will be no need to specify $\boldsymbol{\delta}$ as it will be calculated from $\mathbf{\Gamma}$, and we shall have that for all $t \geq 1$

$$\Pr(C_t = i) = \delta_i \qquad (i = 1, 2, \ldots m).$$

Furthermore, for $t \geq 1$, we shall need to specify model equations for the probabilities

$$\pi_t^i = \Pr(X_t = 1|\mathbf{X}^{(t-1)}, \mathbf{Z}^{(t)}, C_t = i) \qquad (i = 1, 2, \ldots m). \tag{4.6}$$

It follows from Equation (4.6) that

$$\Pr(X_t = 0|\mathbf{X}^{(t-1)}, \mathbf{Z}^{(t)}, C_t = i) = 1 - \pi_t^i. \qquad (i = 1, 2, \ldots m). \tag{4.7}$$

For some purposes it will be useful to define, for $i = 1, 2, \ldots, m$, the probability function

$$p_t^i(x) = \Pr(X_t = x|\mathbf{X}^{(t-1)}, \mathbf{Z}^{(t)}, C_t = i) \qquad (x = 0, 1) \qquad (4.8)$$

and then, combining Equations (4.6) and (4.7), we have for $t \geq 1$

$$p_t^i(x) = \pi_t^i x + (1 - \pi_t^i)(1 - x) \qquad (x = 0, 1; i = 1, 2, \ldots m). \qquad (4.9)$$

Just as in Chapter 3, and specifically as in Equation (3.1), we shall model the π_t^i in terms of a logistic regression, but now with a separate equation for each of the hidden states i. We shall use a model of the form

$$\eta_t^i = \beta_{0i} + \sum_{j=1}^{k} \beta_{ji} W_{tj} \qquad (i = 2, 3, \ldots, m), \qquad (4.10)$$

where

$$\eta_t^i = \ln\left(\frac{\pi_t^i}{1 - \pi_t^i}\right) \qquad (i = 2, 3, \ldots, m), \qquad (4.11)$$

with the inverse relationship

$$\pi_t^i = \frac{\exp(\eta_t^i)}{1 + \exp(\eta_t^i)} \qquad (i = 2, 3, \ldots, m). \qquad (4.12)$$

Thus there are k regressor variables W_j $(j = 1, 2, \ldots, k)$ being used, taking values W_{tj}, some of which may be exogenous variables from the process $\{Z_t\}$ and others endogenous variables constructed from the history of the process $\{X_t\}$, for example the lagged value, X_{t-1}. But whether exogenous or endogenous, as regressor variables they are handled in the same way.[2]

It should be noted that the Equations (4.10)-(4.12) are for the states $i = 2, 3, \ldots, m$ and exclude State 1. The reason for this is that in our application here it turns out to be appropriate to have a state, which we take to be State 1, for which the observed values X_t are always 0, in which case

$$\pi_t^1 = 0 \qquad (4.13)$$

and the logit transformation of Equation (4.11) cannot be used. In our application, State 1 corresponds to the mode of behaviour in which Matthew or Luke is not using the text of Mark at all, specifically in the sense that no words of Mark are retained unchanged, and hence $\pi_t^1 = \Pr(X_t = 1|C_t = 1) = 0$.

Similarly, the logit transformation could also not be used if a hidden state i was needed such that $\pi_t^i = 1$. Such a state too would have to be treated separately. In our application it would correspond to a mode of behaviour in which the text of Mark was being retained exactly word for word. Although

[2]In this and the next chapter we use the term *regressor variable*, where in the previous chapter we used *predictor variable*. The two terms are essentially synonymous.

there are passages in Mark which are retained word for word by Matthew or Luke, these are not very long, and it turns out to be more appropriate to have hidden states where the probability of retention may be large, but not equal to 1.

We are now in a position to draw conclusions regarding the total number of parameters in our HMM . Given m, there are $m(m-1)$ transition probabilities γ_{ij} to be specified. There are m rows of the transition matrix with m numbers in each row, but the row-sums are all necessarily equal to 1, so that, once $m-1$ terms in a row have been specified, the mth term is determined. In addition there are $(m-1)(k+1)$ regression parameters, β_{ji} $(j = 0, 1, \ldots, k; i = 2, 3, \ldots, m)$, in Equation (4.10), which gives a total of $(m-1)(m+k+1)$ free parameters to be fitted, which will be done using the method of maximum likelihood.

4.3 The likelihood function

Suppose that we have a series of T observed values of the binary process, $x_1, x_2, \ldots, x_T$. To fit an HMM to the data, using the method of maximum likelihood, the first step is to find an expression for the likelihood function L_T in a form that can be used as a basis for computation.

First we introduce one additional piece of notation. Let $\mathbf{P}_t(x)$ be the $m \times m$ diagonal matrix,

$$\mathbf{P}_t(x) = \text{diag}(p_t^1(x), p_t^2(x), \ldots, p_t^m(x)) \qquad (x = 0, 1; t = 1, 2, \ldots, T),$$

where the $p_t^i(x)$ are as defined in Equation (4.8) and expressed in Equation (4.9), with the π_t^i modelled as in Equations (4.10)-(4.13).[3]

Proposition 1 *The likelihood function L_T is given by*

$$L_T = \boldsymbol{\delta}\mathbf{P}_1(x_1)\boldsymbol{\Gamma}\mathbf{P}_2(x_2)\boldsymbol{\Gamma}\mathbf{P}_3(x_3)\ldots\boldsymbol{\Gamma}\mathbf{P}_T(x_T)\mathbf{1}'. \tag{4.14}$$

If the initial probability vector for the Markov chain, $\boldsymbol{\delta}$, is the stationary distribution then also

$$L_T = \boldsymbol{\delta}\boldsymbol{\Gamma}\mathbf{P}_1(x_1)\boldsymbol{\Gamma}\mathbf{P}_2(x_2)\boldsymbol{\Gamma}\mathbf{P}_3(x_3)\ldots\boldsymbol{\Gamma}\mathbf{P}_T(x_T)\mathbf{1}'. \tag{4.15}$$

A formal proof of Proposition 1 for the basic HMM model is given on

[3] If lagged values of the series up to lag τ are used as regressor variables in the logistic regression then, in practice, the first τ values of the series will be used to create the initial values of the regressor variables and the model will be fitted to the series from $t = \tau + 1$ onwards. If that is the case then we are, strictly speaking, dealing with a conditional likelihood function and, in the following results, the series used for fitting has been re-indexed to start at $t = 1$.

pp. 37-38 of Zucchini and MacDonald (2009), and only minor modification is needed in the case of our model, where Equations (4.1) and (4.2) are replaced by Equations (4.3) and (4.4). The essentially simple structure of the expressions (4.14) and (4.15) for the likelihood function is exhibited in Corollary 2, which follows immediately.

Corollary 2 *If the matrix* $\mathbf{B}_t$ *is defined by*

$$\mathbf{B}_t = \mathbf{\Gamma P}_t(x_t) \qquad (t = 1, 2, \ldots, T) \tag{4.16}$$

then we may write

$$L_T = \boldsymbol{\delta}\mathbf{P}_1(x_1)\mathbf{B}_2\mathbf{B}_3 \ldots \mathbf{B}_T\mathbf{1}'. \tag{4.17}$$

If the initial probability vector for the Markov chain, $\boldsymbol{\delta}$, *is the stationary distribution then also*

$$L_T = \boldsymbol{\delta}\mathbf{B}_1\mathbf{B}_2\mathbf{B}_3 \ldots \mathbf{B}_T\mathbf{1}'. \tag{4.18}$$

From now on we shall assume that the hidden Markov chain is stationary. An algorithm for computing values of the likelihood function, as expressed in Equation (4.15) or (4.18), will be based upon the following iterative approach.

Define the $1 \times m$ row vectors $\boldsymbol{\alpha}_t$ by

$$\boldsymbol{\alpha}_0 = \boldsymbol{\delta} \tag{4.19}$$

and for $t = 1, 2, \ldots, T$

$$\boldsymbol{\alpha}_t = \boldsymbol{\delta}\mathbf{B}_1\mathbf{B}_2 \ldots \mathbf{B}_t = \boldsymbol{\delta} \prod_{s=1}^{t} \mathbf{B}_s, \tag{4.20}$$

so that, from Equation (4.18),

$$L_T = \boldsymbol{\alpha}_T\mathbf{1}'. \tag{4.21}$$

From Equation (4.20) it follows that for $t = 1, 2, \ldots, T$

$$\boldsymbol{\alpha}_t = \boldsymbol{\alpha}_{t-1}\mathbf{B}_t. \tag{4.22}$$

Thus an iterative approach to calculating a value of the likelihood function calculates the value of $\boldsymbol{\alpha}_0$ using Equation (4.19), then uses the iteration of Equation (4.22) to calculate $\boldsymbol{\alpha}_t$ for $t = 1, 2, \ldots, T$, successively, and finally uses Equation (4.21) to calculate the value of the likelihood function.

4.4 Methods for computation

In this section we address some technical computational issues that arise when we attempt to find maximum likelihood estimates of the model parameters. The R code for the functions that emerge is provided in Appendix A.1.

4.4.1 Scaling the likelihood computation

If we attempt to compute values of the likelihood function using a method based on Equations (4.19), (4.22) and (4.21) then a problem that arises is that the values of the components of the vector $\boldsymbol{\alpha}_t$, which are made up of products of probabilities, become exponentially small as t increases, which eventually results in problems of underflow, so that the values are rounded to zero.

To cope with this problem, we follow the method suggested by Zucchini and MacDonald (2009), pp. 46-47. For $t = 0, 1, 2, \ldots, T$ define the $1 \times m$ vector $\boldsymbol{\phi}_t$ to be a scaled version of $\boldsymbol{\alpha}_t$ such that

$$\boldsymbol{\phi}_t \mathbf{1}' = 1. \tag{4.23}$$

Thus for $t = 0, 1, 2, \ldots, T$

$$\boldsymbol{\phi}_t = \frac{\boldsymbol{\alpha}_t}{w_t}, \tag{4.24}$$

where

$$w_t = \boldsymbol{\alpha}_t \mathbf{1}'. \tag{4.25}$$

In particular, using Equations (4.19) and (4.25),

$$w_0 = \boldsymbol{\alpha}_0 \mathbf{1}' = \boldsymbol{\delta} \mathbf{1}' = 1$$

and, using Equations (4.21) and (4.25),

$$w_T = L_T.$$

Using Equation (4.24) to substitute for $\boldsymbol{\alpha}_t$ into Equation (4.22), we obtain

$$w_t \boldsymbol{\phi}_t = w_{t-1} \boldsymbol{\phi}_{t-1} \mathbf{B}_t. \tag{4.26}$$

From Equation (4.26) we see that for $t = 1, 2, \ldots, T$

$$\boldsymbol{\phi}_t \propto \boldsymbol{\phi}_{t-1} \mathbf{B}_t, \tag{4.27}$$

where the constant of proportionality may be obtained by using Equation (4.23). Post-multiplying Equation (4.26) by $\mathbf{1}'$ and using Equation (4.23), we obtain

$$w_t = w_{t-1} \boldsymbol{\phi}_{t-1} \mathbf{B}_t \mathbf{1}'. \tag{4.28}$$

Taking logarithms in Equation (4.28),

$$\ln w_t = \ln w_{t-1} + \ln \left(\boldsymbol{\phi}_{t-1} \mathbf{B}_t \mathbf{1}' \right). \tag{4.29}$$

Use of the recursions (4.27) and (4.29) provides us with the algorithm for computing the log-likelihood that is presented in Figure 4.1. This algorithm is implemented in Appendix A.1 in the R code for the function `logistic.HMM0.mllk`, which calculates minus the log-likelihood.

$$
\begin{aligned}
&\text{set } \boldsymbol{\phi}_0 \leftarrow \boldsymbol{\delta} \text{ and } \ln w_0 \leftarrow 0 \\
&\text{for } t = 1, 2, \ldots, T \\
&\qquad \mathbf{v} \leftarrow \boldsymbol{\phi}_{t-1} \boldsymbol{\Gamma} \mathbf{P}_t(x_t) \\
&\qquad u \leftarrow \mathbf{v} \mathbf{1}' \\
&\qquad \ln w_t \leftarrow \ln w_{t-1} + \ln u \\
&\qquad \boldsymbol{\phi}_t \leftarrow \mathbf{v}/u \\
&\text{return } \ln w_T \qquad\qquad \text{\# the log-likelihood, } \ln L_T
\end{aligned}
$$

Figure 4.1
Algorithm for calculating the log-likelihood for the HMM

4.4.2 Minimization and the working parameters

Following Zucchini and MacDonald (2009), we shall use the R function `nlm` to minimize minus the log-likelihood with respect to the model parameters to obtain the maximum likelihood estimates. The natural parameters to use are the transition probabilities γ_{ij} $(i = 1, 2, \ldots m, j = 1, 2, \ldots, m)$ as specified in Equation (4.5) and the $(m-1)(k+1)$ regression coefficients β_{ji} $(j = 0, 1, \ldots, k; i = 2, 3, \ldots, m)$ that appear in Equation (4.10).

The function `nlm` numerically carries out an unconstrained minimization, but our natural transition probability parameters satisfy the following constraints:

$$\gamma_{ij} \geq 0 \qquad (i = 1, 2, \ldots m, j = 1, 2, \ldots, m),$$

$$\sum_{j=1}^{m} \gamma_{ij} = 1 \qquad (i = 1, 2, \ldots m),$$

We convert our natural transition probability parameters into working parameters that are unconstrained, so that we may apply the function `nlm` to minimize minus the log-likelihood function as a function of the working parameters, including the regression coefficients, which we leave unchanged. When this has been done, we can convert the resulting working parameters back to natural parameters to obtain the maximum likelihood estimates of the natural parameters.

To correspond to the m^2 natural parameters γ_{ij} $(i = 1, 2, \ldots m, j = 1, 2, \ldots, m)$, assuming that these are all strictly positive, we use the $m(m-1)$ working parameters τ_{ij} $(i = 1, 2, \ldots m, j = 1, 2, \ldots, m, i \neq j)$, defined using an extension of the logit transformation,

$$\tau_{ij} = \ln\left(\frac{\gamma_{ij}}{\gamma_{ii}}\right) \qquad (i = 1, 2, \ldots m, j = 1, 2, \ldots, m, i \neq j). \qquad (4.30)$$

The τ_{ij} are unconstrained. The inverse transformation is given by

$$\gamma_{ii} = \frac{1}{1+\sum_{k\neq i}\exp(\tau_{ik})} \qquad (i=1,2,\dots m), \tag{4.31}$$

$$\gamma_{ij} = \frac{\exp(\tau_{ij})}{1+\sum_{k\neq i}\exp(\tau_{ik})} \qquad (i\neq j). \tag{4.32}$$

The following points are worth noting.

- The total number of working parameters, including the regression coefficients, is

  $$m(m-1)+(m-1)(k+1)=(m-1)(m+k+1),$$

  the same as the total number of free model parameters as described at the end of Section 4.2.

- If we define $\tau_{ii}=0$ so that $\exp(\tau_{ii})=1$ $(i=1,2,\dots,m)$, we may write Equations (4.31) and (4.32) more simply as

  $$\gamma_{ij} = \frac{\exp(\tau_{ij})}{\sum_{k=1}^{m}\exp(\tau_{ik})} \qquad (i=1,2,\dots m, j=1,2,\dots,m). \tag{4.33}$$

- The log-likelihood function is a complicated function of the parameters and may well have several local maxima. There is no guarantee that the iterations of the `nlm` function will converge and, if they do converge, that they will necessarily converge to a global maximum. An appropriate choice of the initial values of the parameters is an important issue, and it will often be sensible to try a number of different sets of initial values.

The function `logistic.HMM0.pn2pw` in Appendix A.1 converts the natural parameters to a vector of working parameters. The function `logistic.HMM0.pw2pn` converts a vector of working parameters to natural parameters and also calculates the stationary distribution for the hidden Markov chain, using the method to be presented in Section 4.4.3. Together with the function `logistic.HMM0.mllk` that calculates minus the log-likelihood, these two functions are used in the function `logistic.HMM0.mle` that computes the maximum likelihood estimates of the model parameters.

4.4.3 Computing a stationary distribution

As stated in Section 4.3, we shall assume that the hidden Markov chain is stationary, which will require the calculation of the stationary distribution from the transition matrix of the Markov chain.

Let $\mathbf{\Gamma}$ be an $m\times m$ transition matrix. As noted earlier, a $1\times m$ vector $\boldsymbol{\delta}$ is said to be a stationary distribution for $\mathbf{\Gamma}$ if it is a solution of

$$\boldsymbol{\delta}\mathbf{\Gamma}=\boldsymbol{\delta}, \tag{4.34}$$

that satisfies the normalization condition

$$\boldsymbol{\delta}\mathbf{1}' = 1. \tag{4.35}$$

Following Zucchini and MacDonald (2009), pp. 26-27, who in turn follow Grimmett and Stirzaker (2001), p. 242, Exercise 6.6.5, we shall use the following result to provide a method of computing the unique stationary distribution for an irreducible, ergodic Markov chain.

Theorem 3 *A vector $\boldsymbol{\delta}$ is a stationary distribution for $\boldsymbol{\Gamma}$ if and only if*

$$\boldsymbol{\delta}(\mathbf{I}_m - \boldsymbol{\Gamma} + \mathbf{U}_m) = \mathbf{1}, \tag{4.36}$$

where $\mathbf{I}_m$ is the $m \times m$ identity matrix and $\mathbf{U}_m$ is the $m \times m$ matrix of 1s.

Proof. Suppose that $\boldsymbol{\delta}$ is a stationary distribution. From Equation (4.34), $\boldsymbol{\delta}(\mathbf{I}_m - \boldsymbol{\Gamma}) = \mathbf{0}$, where $\mathbf{0}$ is the $1 \times m$ vector of 0s. Using the normalization condition (4.35), we find that $\boldsymbol{\delta}\mathbf{U}_m = \mathbf{1}$. Hence Equation (4.36) follows.

Conversely, suppose that Equation (4.36) holds. Note that (i) since the row sums of both $\mathbf{I}_m$ and $\boldsymbol{\Gamma}$ are 1, $\mathbf{I}_m\mathbf{1}' = \boldsymbol{\Gamma}\mathbf{1}' = \mathbf{1}'$ and (ii) from the definition of $\mathbf{U}_m$, $\mathbf{U}_m\mathbf{1}' = m\mathbf{1}'$. Postmultiplying Equation (4.36) by $\mathbf{1}'$, we obtain

$$m\boldsymbol{\delta}\mathbf{1}' = m.$$

Hence $\boldsymbol{\delta}$ satisfies the normalization condition (4.35). It follows that $\boldsymbol{\delta}\mathbf{U}_m = \mathbf{1}$. Hence from Equation (4.36) $\boldsymbol{\delta}(\mathbf{I}_m - \boldsymbol{\Gamma}) = \mathbf{0}$, which yields Equation (4.34). Thus $\boldsymbol{\delta}$ is a stationary distribution.

4.5 Choosing an HMM

An issue that arises when fitting an HMM to an observed time series of length T is, firstly and fundamentally, how many hidden states m should be used and, secondly, how many regressor variables k should be used. In comparing the fitted models for different values of m and k and for different choices of the regressor variables, we may consider choosing a model so as to minimize the Akaike information criterion (AIC),

$$\text{AIC} = -2\ln L_T + 2(m-1)(m+k+1),$$

where, as before, L_T is the likelihood and $(m-1)(m+k+1)$ is the number of free parameters, or to minimize the Bayesian information criterion (BIC),

$$\text{BIC} = -2\ln L_T + (m-1)(m+k+1)\ln T,$$

the second of which tends to suggest models with fewer parameters. However, there will be other considerations to take into account in choosing a model.

One of these will be whether the fitted model can be interpreted in a meaningful way, especially as regards the values of the fitted regression coefficients.

It is, perhaps, worth noting at this stage that we do not have available for these HMM models the theory that would enable us to assign in a straightforward way standard errors or confidence intervals to the fitted transition probabilities and regression coefficients. As discussed by Zucchini and MacDonald (2009), pp. 53-56, for simpler models without regressor variables, it might be possible to use approximate theoretical or bootstrap methods. However, in the presence of regressor variables, such an approach would be a complex and, for bootstrap methods, computationally very time-consuming task. As it was not of critical importance for the way in which the fitted HMMs would be used in the present application, this avenue was left unexplored.

When, given a fitted model, we construct the most likely sequence of hidden states to have given rise to the observed data by the methods of decoding to be described in Section 4.6, what we obtain are sections of the series of varying lengths where the process is in one of the hidden states, before a transition occurs that takes the process into one of the other hidden states. It will be more useful if longer sections of the series with the same hidden state emerge rather than shorter ones, so that, in our particular application, longer sections of text are obtained, perhaps consisting of several verses or the whole of a pericope, where an author is envisaged as working in a particular way, and which may be candidates for detailed textual analysis. Models which represent the author as switching frequently from one mode of behaviour to another will be less useful and perhaps also inherently less plausible in that it seems unlikely that an author would be changing his way of handling the source material at intervals of just a very few words. So we shall prefer models for which the diagonal elements of the transition matrix $\mathbf{\Gamma}$ are closer to 1, especially for states that are of special interest. The other extreme of finding models where the passages identified are too long to be useful does not arise in our context.

As an aid to determining the usefulness of a fitted model in this regard, in Theorem 4 we obtain an expression for the expected length of stay in a hidden state.

Theorem 4 *Given an HMM model with $m \times m$ transition matrix $\mathbf{\Gamma}$, assumed to be irreducible and ergodic, and a corresponding stationary distribution $\boldsymbol{\delta}$, in equilibrium, if a switch to a new hidden state has just occurred, the expected length of stay in that hidden state, including the time of initial entry, is given by*

$$\frac{1}{\sum_{k=1}^{m} \delta_k (1 - \gamma_{kk})} .$$

Proof We first find an expression for the probability in equilibrium that the newly entered hidden state is i.

$$\begin{aligned}
&\Pr(C_{t+1} = i | C_{t+1} \neq C_t) \\
= \quad & \frac{\Pr(C_{t+1} = i, C_{t+1} \neq C_t)}{\Pr(C_{t+1} \neq C_t)} \\
= \quad & \frac{\Pr(\cup_{\{j:j\neq i\}}\{C_t = j, C_{t+1} = i\})}{\Pr(\cup_{k=1}^m \cup_{\{j:j\neq k\}} \{C_t = j, C_{t+1} = k\})} \\
= \quad & \frac{\sum_{\{j:j\neq i\}} \Pr(C_t = j, C_{t+1} = i)}{\sum_{k=1}^m \sum_{\{j:j\neq k\}} \Pr(C_t = j, C_{t+1} = k)} \\
= \quad & \frac{\sum_{\{j:j\neq i\}} \Pr(C_t = j) \Pr(C_{t+1} = i | C_t = j)}{\sum_{k=1}^m \sum_{\{j:j\neq k\}} \Pr(C_t = j) \Pr(C_{t+1} = k | C_t = j)} \\
= \quad & \frac{\sum_{\{j:j\neq i\}} \delta_j \gamma_{ji}}{\sum_{k=1}^m \sum_{\{j:j\neq k\}} \delta_j \gamma_{jk}} \, .
\end{aligned}$$

Now from Equation (4.34) we have that $\sum_{j=1}^m \delta_j \gamma_{ji} = \delta_i$, so that

$$\sum_{\{j:j\neq i\}} \delta_j \gamma_{ji} = \delta_i (1 - \gamma_{ii}).$$

Hence, in equilibrium,

$$\Pr(C_{t+1} = i | C_{t+1} \neq C_t) = \frac{\delta_i (1 - \gamma_{ii})}{\sum_{k=1}^m \delta_k (1 - \gamma_{kk})} \, .$$

Conditional upon state i having been entered, the length of stay in state i, including the time of initial entry, has a geometric distribution with ratio γ_{ii} and expected value $1/(1 - \gamma_{ii})$. Hence, unconditionally, the expected length of stay in the state just entered is given by

$$\begin{aligned}
\sum_{i=1}^m \Pr(C_{t+1} = i | C_{t+1} \neq C_t) \left(\frac{1}{1 - \gamma_{ii}} \right) \quad &= \quad \frac{\sum_{i=1}^m \delta_i}{\sum_{k=1}^m \delta_k (1 - \gamma_{kk})} \\
&= \quad \frac{1}{\sum_{k=1}^m \delta_k (1 - \gamma_{kk})} \, ,
\end{aligned}$$

which is the result of the theorem.

Given any fitted HMM, the expression of Theorem 4 for the expected length of stay in a hidden state, the *mean stay*,

$$\mu = \frac{1}{\sum_{k=1}^m \delta_k (1 - \gamma_{kk})} \, ,$$

may be evaluated, with some preference being given to models with larger values of μ. However, there may be circumstances in which a model with a smaller value of μ is preferred if the value of the expected length of stay $1/(1-\gamma_{ii})$ is greater for an individual state i that is of special interest in providing sections of the series for further investigation.

The function `logistic.HMM0.mle` in Appendix A.1 that computes the maximum likelihood estimates of the model parameters also computes and outputs the resulting values of the AIC, the BIC and the mean stay μ.

In brief, for deciding between different HMM models, the following factors may be taken into account:

1. The goodness of fit of the model as measured by the BIC or AIC, with smaller values preferred

2. The value of the mean stay μ, with larger values preferred

3. The existence of a natural interpretation of the hidden states and model equations

4. The existence of a hidden state or states with model equations whose interpretation is such that it appears possible that decoding may yield for these states sections of the observed series that are potentially of special interest for further analysis

5. After decoding has been carried out, the resulting identification of sections of the observed series that are indeed particularly significant for understanding aspects of its behaviour.

The first two factors are objective in that they are based on numerical values that arise from the model fitting. However they may be in some tension with each other, in that, for example, models that give the smallest value of the BIC or AIC may give undesirably small values of μ. The more subjective factors 3 and 4, that come into play when the fitted models are examined in more detail, may give preference to models that are sub-optimal in the sense of the first two factors but are intuitively more appealing and appear to be more promising in being likely to yield interesting sections of the series for further analysis. Finally, we may be reasonably satisfied with our choice of model if, when the first four factors have been taken into account, the resulting fitted model or models do indeed lead to decoded sections of text that turn out to be of special interest for the problem that we are investigating, in our case especially whether Matthew and Luke appear to be independent in their use of Mark.

4.6 Decoding

Once we have fitted an HMM to our observed binary series, $x_1, x_2, \ldots, x_T$, we may compute what is the most likely sequence of hidden states $c_1, c_2, \ldots, c_T$ to have given rise to the observed series. This construction of the most likely underlying sequence of hidden states is what is known as *global decoding*, as in Zucchini and MacDonald (2009), p. 80. Examination of this sequence of states may provide some insight into the processes that have led to the observed binary series. Indeed, as already remarked in the previous section, if the decoding turns out to be of value for interpreting the observed data, this will provide some justification for the chosen HMM.

Given the observed binary series $\mathbf{x}^{(T)}$ and covariate series $\mathbf{z}^{(T)}$ and a fitted HMM, it is also possible to find for $t = 1, 2, \ldots, T$ a probability distribution for the hidden state C_t,

$$\Pr(C_t = i | \mathbf{X}^{(T)} = \mathbf{x}^{(T)}, \mathbf{Z}^{(T)} = \mathbf{z}^{(T)}) \qquad (i = 1, 2, \ldots, m).$$

For each t we may then choose the value of i that maximizes this probability, which again leads to a sequence of hidden states, $c_1, c_2, \ldots, c_T$. This alternative process of constructing a sequence of hidden states is known as *local decoding*. It should be noted that, although local decoding and global decoding will tend to give similar results, they will usually not give exactly the same sequence of states.

In Section 4.6.1 we provide an outline of the theory behind the method of local decoding and in Section 4.6.2 deal with computational issues. In the results that we present in Chapters 5 and 6, however, we shall focus on global decoding, the method for which is explained in Section 4.6.3.

4.6.1 Forward and backward probabilities and local decoding

For a given fitted model, assume again that the Markov chain of hidden states is stationary, so that the initial vector $\boldsymbol{\delta}$ is the stationary distribution of the Markov chain with transition matrix $\boldsymbol{\Gamma}$. The *forward probabilities* are the $1 \times m$ row vectors $\boldsymbol{\alpha}_t$ as defined in Equations (4.19) and (4.20) in Section 4.3. For convenience we repeat the definitions here:

$$\boldsymbol{\alpha}_0 = \boldsymbol{\delta} \tag{4.37}$$

and for $t = 1, 2, \ldots, T$

$$\boldsymbol{\alpha}_t = \boldsymbol{\delta}\mathbf{B}_1\mathbf{B}_2 \ldots \mathbf{B}_t, \tag{4.38}$$

where $\mathbf{B}_t$ is as defined in Equation (4.16). It follows that for $t = 1, 2, \ldots, T$ the forward probabilities satisfy the recurrence relation

$$\boldsymbol{\alpha}_t = \boldsymbol{\alpha}_{t-1}\mathbf{B}_t. \tag{4.39}$$

Recall also that the likelihood L_T is given by

$$L_T = \boldsymbol{\alpha}_T \mathbf{1}'. \tag{4.40}$$

The *backward probabilities* are the $1 \times m$ row vectors $\boldsymbol{\beta}_t$ defined by

$$\boldsymbol{\beta}_T = \mathbf{1} \tag{4.41}$$

and for $t = 1, 2, \ldots, T-1$

$$\boldsymbol{\beta}_t' = \mathbf{B}_{t+1}\mathbf{B}_{t+2}\ldots\mathbf{B}_T\mathbf{1}'. \tag{4.42}$$

It follows that for $t = 1, 2, \ldots, T-1$ the backward probabilities satisfy the recurrence relation

$$\boldsymbol{\beta}_t' = \mathbf{B}_{t+1}\boldsymbol{\beta}_{t+1}'. \tag{4.43}$$

An explanation of the terminology *forward probabilities* and *backward probabilities* is provided by the results that, for $t = 1, 2, \ldots T$ and $i = 1, 2, \ldots m$,

$$\alpha_t(i) = \Pr(\mathbf{X}^{(t)} = \mathbf{x}^{(t)}, C_t = i | \mathbf{Z}^{(t)} = \mathbf{z}^{(t)}) \tag{4.44}$$

and that for $t = 1, 2, \ldots T-1$ and $i = 1, 2, \ldots m$, provided that $\Pr(C_t = i) > 0$,

$$\beta_t(i) = \Pr(\mathbf{X}_{t+1}^T = \mathbf{x}_{t+1}^T | \mathbf{X}^{(t)} = \mathbf{x}^{(t)}, \mathbf{Z}^{(T)} = \mathbf{z}^{(T)}, C_t = i), \tag{4.45}$$

where $\mathbf{X}_{t+1}^T$ denotes the vector $(X_{t+1}, X_{t+2}, \ldots, X_T)$. Versions of the results of Equations (4.44) and (4.45) are established in Chapter 4 of Zucchini and MacDonald (2009), albeit for a somewhat simpler case, and we shall not give the formal proofs here. Furthermore, Zucchini and MacDonald (2009) also establish slightly simpler versions of the results that we present as Theorem 5 below.

Theorem 5 *For $t = 1, 2, \ldots T$ and $i = 1, 2, \ldots m$,*

(a)

$$\alpha_t(i)\beta_t(i) = \Pr(\mathbf{X}^{(T)} = \mathbf{x}^{(T)}, C_t = i | \mathbf{Z}^{(T)} = \mathbf{z}^{(T)}),$$

(b)

$$L_T = \Pr(\mathbf{X}^{(T)} = \mathbf{x}^{(T)} | \mathbf{Z}^{(T)} = \mathbf{z}^{(T)}) = \boldsymbol{\alpha}_t\boldsymbol{\beta}_t',$$

(c)

$$\Pr(C_t = i | \mathbf{X}^{(T)} = \mathbf{x}^{(T)}, \mathbf{Z}^{(T)} = \mathbf{z}^{(T)}) = \frac{\alpha_t(i)\beta_t(i)}{L_T}.$$

The result of Theorem 5(c) provides the basis for carrying out local decoding. For each $t = 1, 2, \ldots T$, the most likely hidden state is the state i_t for which the probability $\Pr(C_t = i | \mathbf{X}^{(T)} = \mathbf{x}^{(T)}, \mathbf{Z}^{(T)} = \mathbf{z}^{(T)})$ is maximal, i.e.,

$$i_t = \underset{i=1,2,\ldots,m}{\operatorname{argmax}} \Pr(C_t = i | \mathbf{X}^{(T)} = \mathbf{x}^{(T)}, \mathbf{Z}^{(T)} = \mathbf{z}^{(T)}). \tag{4.46}$$

4.6.2 Implementation of the method of local decoding

Functions in R for carrying out local decoding are given in Appendix A.2. The function `logistic.HMM0.lalphabeta` produces matrices that contain the logarithms of the forward and backward probabilities. The function `logistic.HMM0.stateprobs` uses the result of Theorem 5(c) to calculate the probability distribution of the hidden states for each $t = 1, 2, \ldots T$. The function `logistic.HMM0.localdecoding` completes the local decoding by selecting a sequence of states such that for each $t = 1, 2, \ldots T$ the selected state i_t is the one with greatest probability, as in Equation (4.46).

The recursions of Equations (4.39) and (4.43) form the basis for the computational algorithms that are used in the function `logistic.HMM0.lalphabeta`. To avoid numerical problems of underflow, similar scaling methods are used to the one obtained in Section 4.4.1 for computing the log-likelihood. Indeed, the algorithm for computing the logarithms of the forward probabilities, $\ln \boldsymbol{\alpha}_t$, is essentially the same as the one shown in Figure 4.1.

In the case of the backward probabilities $\boldsymbol{\beta}_t$, for $t = 1, 2, \ldots, T$ define the $1 \times m$ row vector $\boldsymbol{\psi}_t$ to be a scaled version of $\boldsymbol{\beta}_t$ such that

$$\boldsymbol{\psi}_t \mathbf{1}' = 1. \tag{4.47}$$

Thus for $t = 1, 2, \ldots, T$

$$\boldsymbol{\psi}_t = \frac{\boldsymbol{\beta}_t}{s_t}\,, \tag{4.48}$$

where

$$s_t = \boldsymbol{\beta}_t \mathbf{1}'. \tag{4.49}$$

In particular, using Equations (4.41) and (4.49),

$$s_T = \boldsymbol{\beta}_T \mathbf{1}' = \mathbf{1}\mathbf{1}' = m.$$

Using Equation (4.48) to substitute for $\boldsymbol{\beta}_{t+1}$ into Equation (4.43), we obtain

$$\boldsymbol{\beta}_t' = s_{t+1} \mathbf{B}_{t+1} \boldsymbol{\psi}_{t+1}', \tag{4.50}$$

so that

$$\ln \boldsymbol{\beta}_t' = \ln(\mathbf{B}_{t+1} \boldsymbol{\psi}_{t+1}') + \ln s_{t+1}. \tag{4.51}$$

Premultiplying Equation (4.50) by $\mathbf{1}$ and using Equation (4.49),

$$s_t = s_{t+1} \mathbf{1} \mathbf{B}_{t+1} \boldsymbol{\psi}_{t+1}'$$

so that

$$\ln s_t = \ln s_{t+1} + \ln(\mathbf{1} \mathbf{B}_{t+1} \boldsymbol{\psi}_{t+1}'). \tag{4.52}$$

Equations (4.51) and (4.52) provide us with an algorithm for computing the logarithms of the backward probabilities, $\ln \boldsymbol{\beta}_t$, as presented in Figure 4.2.

set $\ln \boldsymbol{\beta}_T \leftarrow \mathbf{0}$, $\boldsymbol{\psi}_T \leftarrow \mathbf{1}/m$ and $\ln s_T \leftarrow \ln m$
for $t = T-1, T-2, \ldots, 1$
  $\mathbf{v}' \leftarrow \boldsymbol{\Gamma}\mathbf{P}(x_{t+1})\boldsymbol{\psi}'_{t+1}$
  $\ln \boldsymbol{\beta}_t \leftarrow \ln \mathbf{v} + \ln s_{t+1}$
  $u \leftarrow \mathbf{1}\mathbf{v}'$
  $\boldsymbol{\psi}_t \leftarrow \mathbf{v}/u$
  $\ln s_t \leftarrow \ln s_{t+1} + \ln u$

Figure 4.2
Algorithm for calculating the logarithms of the backward probabilities for the HMM

4.6.3 The Viterbi algorithm and global decoding

In global decoding we search for a sequence of hidden states $c_1, c_2, \ldots, c_T$ that jointly maximizes the conditional probability

$$\Pr(\mathbf{C}^{(T)} = \mathbf{c}^{(T)} | \mathbf{X}^{(T)} = \mathbf{x}^{(T)}, \mathbf{Z}^{(T)} = \mathbf{z}^{(T)}).$$

Equivalently, and more usefully for our purposes, we search for a sequence $c_1, c_2, \ldots, c_T$ that maximizes the joint conditional probability

$$\Pr(\mathbf{C}^{(T)} = \mathbf{c}^{(T)}, \mathbf{X}^{(T)} = \mathbf{x}^{(T)} | \mathbf{Z}^{(T)} = \mathbf{z}^{(T)}).$$

As discussed by Zucchini and MacDonald (2009), pp. 82-84, the Viterbi algorithm is a form of dynamic programming that will carry out this search in an efficient manner. The basis for the method is provided by the result of Theorem 6 below.

Recalling the definition of $p_t^i(x)$ in Equation (4.8) and its expression in Equation (4.9), define

$$\xi_{1i} = \Pr(C_1 = i, X_1 = x_1 | Z_1 = z_1) = \delta_i p_1^i(x_1)$$

and, for $t = 2, 3, \ldots, T$,

$$\xi_{ti} = \max_{c_1, c_2, \ldots, c_{t-1}} \Pr(\mathbf{C}^{(t-1)} = \mathbf{c}^{(t-1)}, C_t = i, \mathbf{X}^{(t)} = \mathbf{x}^{(t)} | \mathbf{Z}^{(t)} = \mathbf{z}^{(t)})$$

So ξ_{ti} is a joint probability up to time t, maximized with respect to $c_1, c_2, \ldots, c_{t-1}$, subject to the constraint that that terminal hidden state at time t is $C_t = i$.

Theorem 6 *For* $t = 2, 3, \ldots, T$ *and* $j = 1, 2, \ldots m$,

$$\xi_{tj} = \left(\max_i (\xi_{t-1,i} \gamma_{ij}) \right) p_t^j(x_t) \ .$$

Proof Abbreviating the notation in order to avoid excessively long expressions,

$$
\begin{aligned}
\xi_{tj} &= \max_{c_1,c_2,\ldots,c_{t-1}} \Pr(\mathbf{c}^{(t-1)}, C_t = j, \mathbf{x}^{(t)} | \mathbf{z}^{(t)}) \\
&= \max_{c_1,c_2,\ldots,c_{t-2},i} \Pr(\mathbf{c}^{(t-2)}, C_{t-1} = i, C_t = j, \mathbf{x}^{(t-1)}, x_t | \mathbf{z}^{(t)}) \\
&= \max_i \left(\max_{c_1,c_2,\ldots,c_{t-2}} \Pr(\mathbf{c}^{(t-2)}, C_{t-1} = i, C_t = j, \mathbf{x}^{(t-1)}, x_t | \mathbf{z}^{(t)}) \right) \\
&= \max_i \left(\max_{c_1,c_2,\ldots,c_{t-2}} \Pr(\mathbf{c}^{(t-2)}, C_{t-1} = i, \mathbf{x}^{(t-1)} | \mathbf{z}^{(t)}) \right. \\
&\qquad \left. \Pr(C_t = j, x_t | \mathbf{c}^{(t-2)}, C_{t-1} = i, \mathbf{x}^{(t-1)}, \mathbf{z}^{(t)}) \right) \\
&= \max_i \left(\max_{c_1,c_2,\ldots,c_{t-2}} \Pr(\mathbf{c}^{(t-2)}, C_{t-1} = i, \mathbf{x}^{(t-1)} | \mathbf{z}^{(t)}) \right. \\
&\qquad \Pr(C_t = j | \mathbf{c}^{(t-2)}, C_{t-1} = i, \mathbf{x}^{(t-1)}, \mathbf{z}^{(t)}) \\
&\qquad \left. \Pr(x_t | \mathbf{c}^{(t-2)}, C_{t-1} = i, C_t = j, \mathbf{x}^{(t-1)}, \mathbf{z}^{(t)}) \right) \\
&= \max_i \left(\max_{c_1,c_2,\ldots,c_{t-2}} \Pr(\mathbf{c}^{(t-2)}, C_{t-1} = i, \mathbf{x}^{(t-1)} | \mathbf{z}^{(t-1)}) \right. \\
&\qquad \left. \Pr(C_t = j | C_{t-1} = i) \Pr(x_t | C_t = j, \mathbf{x}^{(t-1)}, \mathbf{z}^{(t)}) \right) \\
&= \left(\max_i (\xi_{t-1,i} \gamma_{ij}) \right) p_t^j(x_t) \, .
\end{aligned}
$$

Using the result of Theorem 6, we first compute recursively for $t = 1, 2, \ldots, T$ the $T \times m$ matrix of values ξ_{tj}. The required maximizing sequence of hidden states, $c_1, c_2, \ldots, c_T$, is obtained by a backwards recursion:

$$c_T = \operatorname*{argmax}_{i=1,2,\ldots,m} \xi_{Ti} \tag{4.53}$$

and, for $t = T-1, T-2, \ldots, 1$,

$$c_t = \operatorname*{argmax}_{i=1,2,\ldots,m} (\xi_{ti} \gamma_{i,c_{t+1}}). \tag{4.54}$$

This procedure is implemented in the function `logistic.HMM0.viterbi` in Appendix A.2.

4.7 Summary

We have in this chapter outlined the theory of hidden Markov models in a form that is appropriate to our needs for the analysis of the binary time series in Chapters 5 and 6. There are two main aspects to this: firstly, the identification of a suitable model together with the estimation of the parameters, and, secondly, the decoding of the observed series, using the selected model, to uncover the most likely sequence of underlying hidden states. Some of the technical aspects of the corresponding R code, presented in the appendices, have also been explained. This should be quite sufficient information for anyone wishing to replicate our results or to use similar HMMs for the analysis of other binary time series.

The R functions given in the appendices are for the case where State 1 has the special feature that in it the observed values of the series are necessarily 0, so that $\pi_t^1 = 0$. The code for when there is no such special state is very similar and, indeed, somewhat simpler. To avoid repetition, in the appendices we have not listed the code for this simpler alternative, though we have made some use of it in the analysis of Chapter 5. However, the models eventually selected all have $\pi_t^1 = 0$.

There is undoubtedly scope for filling out the details of the theoretical development in a rigorous manner. Furthermore, as mentioned in Section 4.5, we have not attempted to find a method of calculating standard errors for the parameter estimates. This, in particular, is a substantial problem that awaits exploration.

In this monograph the emphasis is more on the application. The use of HMMs will have been justified if the fitted models help us to understand better some aspects of the process of composition of the gospels, and especially if some of the passages obtained from the decoding, and examined in detail in Chapter 7, turn out to be useful in illuminating the nature of the dependence between Matthew and Luke in their verbal agreements with Mark.

5

Matthew's and Luke's use of Mark: hidden Markov models

5.1 Introduction

As already outlined in Section 4.1, we shall now assume that behind the observed binary series (x_t) and (y_t) for Matthew's and Luke's verbal agreements with Mark there lie hidden states that represent different modes of handling the text of Mark. This will lead to the fitting of hidden Markov models (HMMs), for which the technical details were given in Chapter 4. Statistically, the analytical approach taken in this chapter is an alternative to the one taken in Chapter 3, although it shares many of the concepts presented there. We shall consider for use as regressor variables the same variables that were defined in Section 3.3 and used in the logistic regression models of Sections 3.3 and 3.4.

In Section 5.2 we shall suggest what might be the most appropriate HMMs for Matthew's and Luke's use of Mark, respectively, and in Section 5.3 carry out the decoding procedure to find the most likely underlying sequence of hidden states, which will in turn lead to suggested passages of text that might be of particular interest in investigating whether Matthew and Luke are independent in their use of Mark. A first attempt at fitting HMMs to these data is outlined in simpler terms in Abakuks (2015, in press).

5.2 Fitting hidden Markov models

Focussing initially on the series (x_t) that represents Matthew's use of Mark, we suppose that there are m hidden states, which we label $1, 2, 3, \ldots, m$, such that at any point t in the text of Mark the process is in one of these hidden states. For a hidden state i we specify the probability $\pi_t^i = \Pr(X_t = 1)$ along the same lines as in the models of Chapter 3 but with the additional feature, as in Equation (4.6) of Chapter 4, that now the probability depends not only on various regressor variables but also on the hidden state i. Thus, as in Equation

Table 5.1
Potential regressor variables for HMMs for the series (x_t)

Notation	Description
X_{t-1}	the value at lag 1 of the series (x_t)
X_{t-2}	the value at lag 2 of the series (x_t)
R^0_{t-1}	a "log-run length" of 0s of the series (x_t)
R^1_{t-1}	a "log-run length" of 1s of the series (x_t)
Z_t	the current value of the variable for direct speech
Y_t	the current value of the series (y_t)
Y_{t-1}	the value at lag 1 of the series (y_t)
Y_{t-2}	the value at lag 2 of the series (y_t)
S^0_{t-1}	a "log-run length" of 0s of the series (y_t)
S^1_{t-1}	a "log-run length" of 1s of the series (y_t)

(4.10), we have the logistic regression model

$$\ln\left(\frac{\pi^i_t}{1-\pi^i_t}\right) = \beta_{0i} + \sum_{j=1}^{k} \beta_{ji} W_{tj}, \tag{5.1}$$

where the k regressor variables W_j $(j = 1, 2, \ldots, k)$ on the right-hand side of Equation (5.1), taking values W_{tj}, are potentially to be chosen from among the regressor variables considered in Chapter 3, which for reference we now present in Table 5.1.

Given a selection of k regressor variables, if Equation (5.1) is used for all m hidden states, this will give $m(k+1)$ regression parameters to be estimated. In addition, there will be $m(m-1)$ free parameters to be estimated for the $m \times m$ transition matrix for the hidden states, giving a total of $m(m+k)$ free parameters.

We shall not want to have too many hidden states and regressor variables because of the practical numerical problems of estimating large numbers of parameter values and the problems of interpretation of the parameter values that may arise. The hidden states may be expected to reflect whether we are currently in a phase of the text where Matthew is following Mark fairly closely or one where he is following Mark only very loosely or not at all, features which were to some extent represented by the values of the "log-run length" variables, R^0_{t-1}, R^1_{t-1} for Matthew, and S^0_{t-1}, S^1_{t-1} for Luke, that played an important role as regressor variables in the logistic regression analyses of Chapter 3. This suggests that the hidden states may take over some of the role of these variables and perhaps make them redundant. In other words, in the present setting, the regressor variables that are selected may be expected to represent local effects, whereas the hidden states represent the longer term effects of how Matthew is currently handling the text of Mark.

So, although, in our explorations of what might be suitable HMMs to use, we shall consider $R^0_{t-1}, R^1_{t-1}, S^0_{t-1}, S^1_{t-1}$ as potential regressor variables, we may envisage, at least as a possibility, that models without them will be more readily interpretable and, after decoding, more fruitful in finding significant passages of text for further analysis, as discussed in more general terms in Section 4.5.

It is a possibility that was anticipated in the theoretical treatment in Chapter 4, and expressed in the R code of the appendices, that one of the hidden states corresponds to the text of Mark not being used by Matthew at all. Labelling this to be State 1, it follows that necessarily $X_t = 0$ in State 1 and hence

$$\pi^1_t = 0.$$

We shall see that this variant of the model does turn out to be appropriate. The π^i_t for the remaining states $2, 3, \ldots m$ are still specified by Equation (5.1), and, as noted in Section 4.2, there are then $(m-1)(m+k+1)$ free parameters to be fitted.

There is potentially a very large number of different HMMs that can be considered for different choices of the number of hidden states m and for different choices of the regressor variables. In Table 5.2 we present a selection of the more plausible models that arose in our search for what might be the most appropriate model to use. It turned out that $m = 3$ appeared to be a good candidate for the best number of hidden states to use, together with the specification that $\pi^1_t = 0$, represented by a tick in the third column of Table 5.2, and we begin by considering this collection of models for different numbers and selections of the regressor variables. First we consider $k = 1$, taking in turn each of the variables $X_{t-1}, R^0_{t-1}, R^1_{t-1}, Z_t, Y_t$, considered individually. In the logistic regression models of Section 3.4 for modelling (x_t), the best single variable to use turned out to be R^0_{t-1}, but in the present HMM context this is no longer the case. Instead, very clearly, from examination of $-\ln L_T$, or the AIC or BIC, the best single variable to use in terms of goodness of fit is Y_t, so that immediately there is an indication of an association between whether a word is retained unchanged by Matthew and whether it is retained unchanged by Luke. Although the lagged variable X_{t-1} is not the best one to use in terms of goodness of fit, it stands out as the single variable that gives much the largest value for mean stay μ, which, as discussed in Section 4.5, is another important property for an HMM, in that it may be expected to give longer stretches of text with the same hidden state. When a second regressor variable is included in addition to Y_t, so that $k = 2$, the best fit is given by the pair R^0_{t-1}, Y_t, followed by the pair X_{t-1}, Y_t, but the latter has the advantage that it gives a mean stay of 10.62 as against 3.57 for the former.

For the case $m = 3, k = 3$, in Table 5.2 we show only the best two choices of three regressor variables, which are obtained by introducing the variable Z_t for speech, so that we have the triples R^0_{t-1}, Z_t, Y_t and X_{t-1}, Z_t, Y_t. These two sets of three regressor variables overall provide the best models found according to the BIC criterion, with the values highlighted in bold in the

Table 5.2
Comparison of fitted HMMs for Matthew's use of Mark, (X_t)

m	k	$\pi_t^1 = 0$	free/working parameters	regressor variables	$-\ln L_T$	AIC	BIC	mean stay μ	model chosen
2	3	–	10	X_{t-1}, Z_t, Y_t	5567.94	11155.87	11229.00	37.71	
2	3	–	10	R^0_{t-1}, Z_t, Y_t	5543.40	11106.80	11179.93	5.49	
2	3	✓	6	X_{t-1}, Z_t, Y_t	5608.75	11229.50	11273.38	35.57	
2	3	✓	6	R^0_{t-1}, Z_t, Y_t	5606.70	11225.41	11269.28	5.81	
3	0	✓	8		5842.36	11700.71	11759.21	5.26	
3	1	✓	10	X_{t-1}	5819.94	11659.88	11733.01	21.39	
3	1	✓	10	R^0_{t-1}	5817.01	11654.02	11727.14	6.54	
3	1	✓	10	R^1_{t-1}	5832.39	11684.78	11757.90	4.66	
3	1	✓	10	Z_t	5796.78	11613.56	11686.68	5.11	
3	1	✓	10	Y_t	5575.94	11171.87	11245.00	5.69	
3	2	✓	12	X_{t-1}, Y_t	5518.81	11061.62	11149.37	10.62	
3	2	✓	12	R^0_{t-1}, Y_t	5511.45	11046.90	11134.65	3.57	
3	2	✓	12	R^1_{t-1}, Y_t	5524.43	11072.87	11160.62	5.40	
3	2	✓	12	Z_t, Y_t	5538.83	11101.65	11189.40	5.80	
3	3	✓	14	X_{t-1}, Z_t, Y_t	5483.17	10994.34	**11096.71**	10.11	✓
3	3	✓	14	R^0_{t-1}, Z_t, Y_t	5479.91	10987.81	**11090.19**	4.51	
3	3	–	18	X_{t-1}, Z_t, Y_t	5482.65	11001.31	11132.93	10.04	
3	3	–	18	R^0_{t-1}, Z_t, Y_t	5469.82	10975.64	11107.26	5.07	
3	4	✓	16	$X_{t-1}, X_{t-2}, Z_t, Y_t$	5478.38	10988.76	11105.76	11.23	
3	4	✓	16	$X_{t-1}, R^0_{t-1}, Z_t, Y_t$	5480.87	10993.74	11110.75	10.00	
4	3	✓	24	X_{t-1}, Z_t, Y_t	5454.10	**10956.21**	11131.71	6.65	✓
4	4	✓	27	$X_{t-1}, X_{t-2}, Z_t, Y_t$	5439.76	**10933.52**	11130.96	10.09	

table. The first of these triples provides the best fit in that it gives a higher value of the likelihood and hence a lower value of the BIC, but the latter has the advantage of having a mean stay of 10.11 as against 4.51 for the former. The triple X_{t-1}, Z_t, Y_t was adopted as the model of choice in the expectation that it would be more likely to provide longer passages of text with the same hidden state, suitable for detailed textual analysis.

The last six lines of Table 5.2 show some examples of the better fitting models with more parameters, obtained either by dropping the specification that $\pi_t^1 = 0$, or by adding a further regressor variable, or by adding a fourth hidden state. All these cases lead to a larger value of the BIC. They also lead to other complications in that the regression coefficients and the hidden states may not have simple, natural interpretations. Furthermore, with an increase in the number of parameters there is a greater tendency for problems to arise with the convergence of the maximum likelihood procedure, and the solution arrived at may be very sensitive to the choice of initial parameter values, so that there may be some doubt as to whether a global maximum of the likelihood function really has been found. Nevertheless, the last two models, with four hidden states, $m = 4$, and with $k = 3$ regressor variables X_{t-1}, Z_t, Y_t and $k = 4$ regressor variables $X_{t-1}, X_{t-2}, Z_t, Y_t$, respectively, are the ones with smallest AIC overall. The second of these, with the smaller AIC and the larger μ, however, turned out to be less satisfactory than the first in that, when decoding was carried out, the State 2 passages, which, as we shall see later, turned out to be the most promising ones for further investigation, were for this model rather too short to be useful. This could be anticipated from examination of the fitted transition matrix, which gave a relatively small value of $\gamma_{22} = 0.781$ and hence of the expected length of stay in State 2, $1/(1 - \gamma_{22}) = 4.56$. Hence, as an alternative to the chosen three-state model, the four-state model with the same set of regressor variables X_{t-1}, Z_t, Y_t was chosen.

The first four lines of Table 5.2 show models with only two hidden states, i.e., $m = 2$, but using the same two sets of three regressor variables that with $m = 3$ gave the lowest values of the BIC. If the specification that $\pi_t^1 = 0$ is adopted, this gives a model with one state where Mark is not being used at all by Matthew and leaves the other state to correspond to the text of Mark being used by Matthew, without having a means through the hidden state structure to represent different types of use of the text of Mark. It is then perhaps not surprising that such models give a relatively poor fit. With $m = 2$ the models without the specification that $\pi_t^1 = 0$ do much better according to the BIC criterion, but still much worse than the corresponding models with $m = 3$.

Below we give the fitted model for the chosen number of hidden states, $m = 3$, and the chosen regressor variables X_{t-1}, Z_t, Y_t. Using the methods of Chapter 4, the fitted transition matrix $\mathbf{\Gamma}$ for the hidden states is given by

$$\mathbf{\Gamma} = \begin{pmatrix} 0.9284 & 0.0669 & 0.0046 \\ 0.0322 & 0.8488 & 0.1190 \\ 0.0096 & 0.0677 & 0.9227 \end{pmatrix} . \tag{5.2}$$

The fitted model equations are

$$\pi_t^1 = 0 \tag{5.3}$$

$$\ln\left(\frac{\pi_t^2}{1-\pi_t^2}\right) = -2.193 + 0.567X_{t-1} + 0.131Z_t + 4.099Y_t \tag{5.4}$$

$$\ln\left(\frac{\pi_t^3}{1-\pi_t^3}\right) = -0.542 + 1.523X_{t-1} + 0.754Z_t + 0.003Y_t \tag{5.5}$$

The signs of the coefficients on the right-hand sides of Equations (5.4) and (5.5) show that in both State 2 and State 3 a word in Mark is more likely to be retained unchanged by Matthew if the previous word was retained by Matthew, if the current word is part of the direct speech of Jesus, and if the current word was retained by Luke. Comparing the sizes of the coefficients, especially for the constant term, we observe that State 2 appears to represent overall a less intensive use of Mark by Matthew than does State 3, which might be taken to suggest substantial influence from oral tradition, but with the important qualification that the large coefficient 4.099 for Y_t in Equation (5.4) indicates that in State 2 there is a particularly strong positive association between whether a word in Mark is retained by Matthew and whether it is retained by Luke. So it may be anticipated that, after decoding, the State 2 passages might provide examples of texts with strong evidence of dependence between Matthew and Luke in their use of Mark.

Although we tend to prefer the above HMM with three hidden states, as a plausible alternative we give below the fitted model for the HMM with the same regressor variables but with four hidden states, $m = 4$, that would be preferable according to the AIC criterion. The fitted transition matrix $\mathbf{\Gamma}$ for the hidden states is given by

$$\mathbf{\Gamma} = \begin{pmatrix} 0.9349 & 0.0555 & 0.0016 & 0.0079 \\ 0.0278 & 0.8747 & 0.0529 & 0.0446 \\ 0.0029 & 0.0456 & 0.7250 & 0.2265 \\ 0.0128 & 0.0321 & 0.1024 & 0.8527 \end{pmatrix}. \tag{5.6}$$

The fitted model equations are

$$\pi_t^1 = 0 \tag{5.7}$$

$$\ln\left(\frac{\pi_t^2}{1-\pi_t^2}\right) = -2.724 + 0.787X_{t-1} + 0.663Z_t + 4.304Y_t \tag{5.8}$$

$$\ln\left(\frac{\pi_t^3}{1-\pi_t^3}\right) = 1.550 + 0.600X_{t-1} - 3.471Z_t + 0.792Y_t \tag{5.9}$$

$$\ln\left(\frac{\pi_t^4}{1-\pi_t^4}\right) = -1.198 + 1.023X_{t-1} + 3.112Z_t + 0.603Y_t \tag{5.10}$$

In this case too we see that State 2 appears to be one in which there is a strong positive association between whether a word in Mark is retained unchanged by Matthew and whether it is retained unchanged by Luke, as shown by the coefficients in Equation (5.8).

A noteworthy feature of Equation (5.9) is the positive constant term on the right-hand side, which indicates that on the whole, in State 3, Matthew follows Mark quite closely; but what is surprising is the large negative coefficient for Z_t, which suggests a strong negative association between whether a word in Mark is retained unchanged by Matthew and whether it is part of the direct speech of Jesus. However, the value of $\gamma_{33} = 0.725$ is small, and hence so is the expected length of stay in State 3, $1/(1 - \gamma_{33}) = 3.64$. Anticipating the results of the decoding in Section 5.3, we note that the State 3 passages tend to be very short, and those that are more than a very few words in length turn out to be examples of short pieces of narrative material, with no speech of Jesus, but where there is quite close verbal agreement between Matthew and Mark. These comments are somewhat tangential to our main focus on State 2 passages, but they lend support our earlier comments that the more parameters are introduced into our model, the greater the likelihood that problems of interpretation and perhaps spurious effects will occur.

We now turn to the series (y_t) for Luke's use of Mark. We use the notation that, for a hidden state i, $\theta_t^i = \Pr(Y_t = 1)$ and let $\boldsymbol{\Delta} = (\delta_{ij})$ denote the transition matrix for the hidden states. In Table 5.3 we present a selection of some of the better fitting HMMs, with regressor variables chosen from among those in Table 5.1, but with Y_t replaced as a regressor variable by X_t. The choice between models here is rather more difficult than was the case for Matthew's use of Mark. It turns out that in modelling Luke's use of Mark there is a stronger argument for using four hidden states instead of three than was the case in modelling Matthew's use of Mark. We have chosen to present for discussion in Table 5.3 mainly examples with four hidden states ($m = 4$), and in all these cases it is definitely better to specify State 1 to be such that $\theta_t^1 = 0$. Lengthy passages such that Mark's text is not being used at all are much more emphatically present for Luke's use of Mark than was the case for Matthew's use of Mark, notably Luke's great omission of Mk 6:45-8:10, and such passages will be modelled by State 1.

We begin by considering HMMs for $m = 4$ and different numbers and selections of the regressor variables. First we consider $k = 1$, taking in turn each of the variables $Y_{t-1}, S_{t-1}^0, S_{t-1}^1, Z_t, X_t$, considered individually. In a way that parallels what we found for Matthew's use of Mark, from examination of $-\ln L_T$, or the AIC or BIC, the best single variable to use in terms of goodness of fit is X_t, so that immediately there is again a strong indication of an association between whether a word is retained unchanged by Matthew and whether it is retained unchanged by Luke. Although the lagged variable Y_{t-1} is not the best one to use in terms of goodness of fit, it stands out as the single variable that gives much the largest value for the mean stay μ, which may be expected to lead to longer stretches of text with the same hidden

Table 5.3
Comparison of fitted HMMs for Luke's use of Mark, (X_t)

m	k	$\theta_t^1 = 0$	free/working parameters	regressor variables	$-\ln L_T$	AIC	BIC	mean stay μ	model chosen
3	3	✓	14	Y_{t-1}, Z_t, X_t	4309.99	8647.98	8750.36	8.76	✓
3	3	✓	14	Y_{t-1}, S^0_{t-1}, X_t	4295.38	8618.77	8721.14	6.19	
3	3	✓	14	$S^0_{t-1}, S^1_{t-1}, X_t$	4264.60	8557.19	**8659.57**	5.76	
3	3	✓	14	S^0_{t-1}, Z_t, X_t	4298.49	8624.98	8727.35	3.71	
3	4	✓	16	$Y_{t-1}, S^0_{t-1}, Z_t, X_t$	4280.68	8593.36	8710.36	6.44	
3	4	✓	16	$S^0_{t-1}, S^1_{t-1}, Z_t, X_t$	4254.79	8541.58	**8658.58**	6.01	
4	0	✓	15		4615.60	9261.20	9370.88	5.84	
4	1	✓	18	Y_{t-1}	4596.88	9229.77	9361.39	10.13	
4	1	✓	18	S^0_{t-1}	4595.40	9226.79	9358.42	2.15	
4	1	✓	18	S^1_{t-1}	4597.25	9230.49	9362.12	3.96	
4	1	✓	18	Z_t	4580.03	9196.06	9327.69	6.21	
4	1	✓	18	X_t	4298.75	8633.49	8765.12	5.97	
4	2	✓	21	Y_{t-1}, X_t	4280.99	8603.98	8757.54	8.85	
4	2	✓	21	S^0_{t-1}, X_t	4263.24	8568.49	8722.05	4.71	
4	2	✓	21	S^1_{t-1}, X_t	4272.66	8587.32	8740.89	5.55	
4	2	✓	21	Z_t, X_t	4284.90	8611.80	8765.36	6.10	
4	3	✓	24	Y_{t-1}, Z_t, X_t	4250.68	8549.37	8724.87	10.04	✓
4	3	✓	24	$S^0_{t-1}, S^1_{t-1}, X_t$	4252.42	8552.84	8728.34	4.84	
4	3	✓	24	S^0_{t-1}, Z_t, X_t	4260.33	8568.67	8744.17	7.22	
4	4	✓	27	$Y_{t-1}, S^0_{t-1}, Z_t, X_t$	4232.85	**8519.70**	8717.14	9.81	
4	4	✓	27	$S^0_{t-1}, S^1_{t-1}, Z_t, X_t$	4231.80	**8517.60**	8715.03	4.88	

state. When a second regressor variable is included in addition to X_t, so that $k = 2$, the best fit is given by the pair S^0_{t-1}, X_t, followed by the pair S^1_{t-1}, X_t, but both these pairs have rather small values of μ. Although the pair Y_{t-1}, X_t gives only the third best fit in the sense of maximum likelihood, it has a larger value of $\mu = 8.85$.

For the case $m = 4, k = 3$, the best fit is given by the triple of regressor variables Y_{t-1}, Z_t, X_t, which also has the highest value of $\mu = 10.04$. This then appears to be a very attractive model to use, especially as it also avoids use of the run-length variables S^0_{t-1} and S^1_{t-1}, whose role of representing longer term effects we have earlier anticipated might be taken over by the hidden states.

The last two lines of Table 5.3 show the two best fitting models with $m = 4, k = 4$. These give the smallest values of the AIC, but the second of these with regressor variables $S^0_{t-1}, S^1_{t-1}, Z_t, X_t$ has a small value of the mean stay μ, and decoding will not provide useful passages of text for analysis. The model with regressor variables $Y_{t-1}, S^0_{t-1}, Z_t, X_t$ has a much higher value of μ, which suggests that decoding might provide potentially useful sections of text for detailed investigation. However, this model is difficult to interpret. A number of the signs of the coefficients are not what would be expected, and it was decided to give preference to the model found in the previous paragraph with $m = 4, k = 3$ and regressor variables Y_{t-1}, Z_t, X_t, which presents fewer problems of interpretation and which parallels the model chosen for Matthew's use of Mark.

The first six lines of Table 5.3 show the best fitting models with only three hidden states, $m = 3$. The ones that give the smallest values of the BIC are the model with $k = 4$ and regressor variables $S^0_{t-1}, S^1_{t-1}, Z_t, X_t$ and the model with $k = 3$ and regressor variables $S^0_{t-1}, S^1_{t-1}, X_t$. But these models have relatively small values of the mean stay μ, so that they will not provide useful passages of text for analysis. Furthermore, State 2 turns out to be the one of special interest, and for both these models the fitted values of δ_{22} and hence of the expected length of stay in State 2, $1/(1 - \delta_{22})$, turn out to be particularly small. Instead, we choose again to work with the model $k = 3$ and regressor variables Y_{t-1}, Z_t, X_t, which has a larger value of the mean stay μ.

In summary, then, for Luke's use of Mark, for the reasons given above we do not choose the models that are suggested by a simple application of either the BIC or the AIC criterion. Instead, in a way that parallels the models that we chose for Matthew's use of Mark, we choose the models with $k = 3$ and regressor variables Y_{t-1}, Z_t, X_t and with either three hidden states ($m = 3$) or four hidden states ($m = 4$). However, here, for Luke's use of Mark, whether by the AIC, BIC or mean stay criterion, the model with $m = 4$ is preferred. We begin by presenting the fitted model for $m = 3$. The fitted transition matrix $\mathbf{\Delta}$ for the hidden states is given by

$$\mathbf{\Delta} = \begin{pmatrix} 0.9737 & 0.0186 & 0.0078 \\ 0.0242 & 0.8007 & 0.1751 \\ 0.0070 & 0.1273 & 0.8657 \end{pmatrix}. \tag{5.11}$$

The fitted model equations are

$$\theta_t^1 = 0 \tag{5.12}$$

$$\ln\left(\frac{\theta_t^2}{1-\theta_t^2}\right) = -3.168 + 1.015Y_{t-1} + 0.652Z_t + 4.603X_t \tag{5.13}$$

$$\ln\left(\frac{\theta_t^3}{1-\theta_t^3}\right) = -1.242 + 1.996Y_{t-1} + 0.395Z_t - 0.343X_t \tag{5.14}$$

The signs of the coefficients on the right-hand sides of Equations (5.13) and (5.14) show that in both State 2 and State 3 a word in Mark is more likely to be retained unchanged by Luke if the previous word was retained by Luke, especially so in State 3, and if the current word is part of the direct speech of Jesus. State 2 appears to represent a generally less intensive use of Mark by Luke than does State 3, but the especially large coefficient 4.603 for X_t in Equation (5.13) indicates, as in Equation (5.4) for Matthew's use of Mark, that in State 2 there is a particularly strong positive association between whether a word in Mark is retained by Matthew and whether it is retained by Luke. In Equation (5.14) the negative coefficient for X_t shows that in State 3 for Luke's use of Mark, a word is less likely to be retained by Luke if it is retained by Matthew, which is a somewhat surprising negative association.

The fitted model for $m = 4$ has transition matrix

$$\mathbf{\Delta} = \begin{pmatrix} 0.9804 & 0.0160 & 0.0035 & 0.0001 \\ 0.0193 & 0.8582 & 0.0675 & 0.0550 \\ 0.0000 & 0.1113 & 0.8685 & 0.0202 \\ 0.0001 & 0.0714 & 0.0495 & 0.8791 \end{pmatrix}. \tag{5.15}$$

The fitted model equations are

$$\theta_t^1 = 0 \tag{5.16}$$

$$\ln\left(\frac{\theta_t^2}{1-\theta_t^2}\right) = -4.447 + 1.035Y_{t-1} + 1.652Z_t + 3.877X_t \tag{5.17}$$

$$\ln\left(\frac{\theta_t^3}{1-\theta_t^3}\right) = -0.959 + 1.858Y_{t-1} - 0.618Z_t - 0.850X_t \tag{5.18}$$

$$\ln\left(\frac{\theta_t^4}{1-\theta_t^4}\right) = -1.168 + 1.113Y_{t-1} + 2.037Z_t + 3.168X_t \tag{5.19}$$

In this case examination of the estimated regression coefficients shows that in

State 4 as well as in State 2 there appears to be a strong positive association between whether a word in Mark is retained unchanged by Matthew and whether it is retained unchanged by Luke, as shown by the coefficients for X_t in Equations (5.17) and (5.19), but on the whole in State 4 Luke is following Mark much more closely than in State 2, where the overall retention rate is so small that a strong influence from oral tradition might be envisaged. A noteworthy feature of Equation (5.18), for State 3, is the emergence of negative coefficients for Z_t and X_t, so that a word in Mark is less likely to be retained unchanged by Luke if it is part of the direct speech of Jesus or if it is retained unchanged by Matthew.

But what is important for our purposes here is that, for this model, it may be anticipated that, after decoding, both State 2 passages and State 4 passages might provide examples of texts with strong evidence of dependence between Matthew's and Luke's use of Mark.

5.3 Decoding the text of Mark for the fitted HMMs

In this section we report on the results of the process of global decoding, as described in general terms in Section 4.6.3, for each of the four models chosen in Section 5.2, principally in order to identify passages that are potentially of interest for investigating any apparent dependence between Matthew and Luke in their use of Mark.

We begin with the fitted HMM with three hidden states for Matthew's use of Mark, as specified in Equations (5.2)-(5.5). As a result of the decoding, we obtain a sequence of State 1, State 2 and State 3 passages of widely varying lengths. For example, the longest State 1 passages, where Matthew is not retaining any words from Mark, are Mk 8:20:4-8:27:1 (97 words) that includes the pericope of the Blind Man of Bethsaida, which is absent from both Matthew and Luke, and Mk 12:40:1-12:44:21 (89 words), the pericope of the Widow's Gift, which is absent from Matthew, though present in Luke. The longest State 3 passage is Mk 15:27:13-15:42:4 (219 words), which consists of part of the pericope of the Crucifixion and the pericope of the Death on the Cross, where Matthew is on the whole following the text of Mark quite closely, but Luke is following Mark much less closely.

However, our main focus for more detailed study is on the State 2 passages, because we anticipate that there we may find the strongest evidence of dependence between Matthew and Luke in their use of Mark. When the decoding is carried out, there are 166 State 2 passages for Matthew's use of Mark with a total of 2773 words, varying in length from 1 to 61 words and with a median of 13.5. If we now consider which of these passages might merit closer examination, we may reject the shorter passages as being unlikely to provide sufficient material for any useful discussion. On the other hand, it may be sensible to

Table 5.4
Longer State 2 passages for Matthew's use of Mark (3 hidden states)

Passage in Mark	No. of words	Amalgamated and rounded	Description of passage
1:39:7-1:44:1	54	Mk 1:40-44	The Healing
	19	(in Huck 45)	of a Leper
1:44:21-1:44:25	5		
3:28:4-3:28:18	15	Mk 3:28-33	A House Divided
	9	(in Huck 86,89)	
3:29:10-3:33:4	60		Jesus' True Relatives
6:37:15-6:41:15	58	Mk 6:37-44	The Feeding of
	5	(in Huck 112)	the Five Thousand
6:41:21-6:44:8	34		

amalgamate longer State 2 passages with neighbouring State 2 passages which are separated from them by only a few words with a different hidden state or states. We may ignore passages which are not used by Luke, as in Luke's great omission, because in such passages Luke is simply not using Mark at all and so we will not be able to detect any correlation between Matthew's and Luke's use of Mark. In Table 5.4 there are presented the State 2 passages for Matthew's use of Mark which exceed 50 words in length, but excluding those which are contained entirely within State 1 passages for Luke's use of Mark. Shorter State 2 passages that are near neighbours are also included and merged. This results in a total of 3 passages for consideration, which are then rounded so that only whole verses are included in a passage. In Table 5.4 the first column gives a precise specification of the State 2 passages. The second column gives the exact number of words in each passage before amalgamation and rounding, and also includes the number of words in intervening passages. The third column gives the specification of the amalgamated and rounded passage, which appears in its own box in the table together with the number or numbers of the Huck pericopes[1] in which it is contained, and the fourth column gives a description of the passage.

When global decoding is carried out for the fitted HMM with four hidden states for Matthew's use of Mark, as specified in Equations (5.6)-(5.10), there are 120 State 2 passages with a total of 2709 words, varying in length from 1 to 86 words and with a median of 19.5. The results are processed and presented in Table 5.5 in a similar fashion to the results for the model with three hidden

[1]See Huck (1949) or Throckmorton (1992) for specification of the pericopes.

Table 5.5
Longer State 2 passages for Matthew's use of Mark (4 hidden states)

Passage in Mark	No. of words	Amalgamated and rounded	Description of passage
1:39:7-1:44:14	67	Mk 1:40-44	The Healing
	6	(in Huck 45)	of a Leper
1:44:21-1:44:25	5		
3:28:4-3:28:18	15	Mk 3:28-33	A House Divided
	9	(in Huck 86,89)	
3:29:10-3:33:4	60		Jesus' True Relatives
5:11:1-5:11:2	2	Mk 5:11-14	The Gerasene
	4	(in Huck 106)	Demoniac
5:11:7-5:14:10	57		
6:37:15-6:41:15	58	Mk 6:37-44	The Feeding of
	5	(in Huck 112)	the Five Thousand
6:41:21-6:45:2	36		
9:33:3-9:37:13	71	Mk 9:33-37	The Dispute about
	7	(Huck 129)	Greatness
9:37:21-9:37:25	5		
15:23:9-15:29:1	50	Mk 15:23-27 (in Huck 249)	The Crucifixion

states.[2] We may note that the rounded passages obtained in Table 5.4 are a subset of those in Table 5.5.

We now consider the results from the global decoding of the fitted HMM with three hidden states for Luke's use of Mark, as specified in Equations (5.11)-(5.14). One of the outstanding features is the presence of exceptionally long State 1 passages, where Luke is not retaining any of the text of Mark. The three longest such passages are Mk 6:45:1-8:11:9 (969 words), which contains Luke's great omission, Mk 6:17:22-6:29:16 (227 words), almost the whole of the pericope of the Death of John the Baptist, and Mk 8:15:12-8:26:12 (158 words), which includes most of the pericope of the Yeast of the Pharisees and of Herod and the pericope of the Blind Man of Bethsaida. The longest State 3 passage is Mk 3:1:7-3:8:22 (132 words), which contains most of the pericope of the Healing of the Man with the Withered Hand.

However, just as for Matthew's use of Mark, our main focus is on the State 2 passages, where we hope to find the strongest evidence of dependence between Matthew and Luke in their use of Mark. The global decoding gives 208 State 2 passages for Luke's use of Mark with a total of 2634 words, varying

[2] With reference to the final passage in Table 5.5, note that the verse Mk 15:28 is omitted in critical editions of the Greek text and in modern translations, as not belonging to the original text of Mark.

Table 5.6
Longer State 2 passages for Luke's use of Mark (3 hidden states)

Passage in Mark	No. of words	Amalgamated and rounded	Description of passage
2:8:4-2:10:15	54	Mk 2:8-12	The Healing of
	12	(in Huck 52)	the Paralytic
2:11:10-2:12:18	22		
12:36:1-12:38:7	51	Mk 12:36-38 (in Huck 209, 210)	About David's Son, Jesus Denounces Scribes

in length from 1 to 54 words and with a median of 10. There are just two State 2 passages which exceed 50 words in length. The results are processed and presented in Table 5.6 as was done in Tables 5.4 and 5.5 for Matthew's use of Mark.

When the global decoding is carried out for the fitted HMM with four hidden states for Luke's use of Mark as specified in Equations (5.15)-(5.19), following on from the earlier discussion of these equations, we consider both the State 2 and State 4 passages. There turn out to be far more longer passages than in the previous cases, and, to avoid an overabundance of passages for consideration, in Table 5.7 there are listed the State 2 and State 4 passages with at least 80 words, with amalgamation of some neighbouring State 2 and State 4 passages and with rounding. This table includes an additional column which specifies the hidden state of any passage. The situation is now more complex than when only State 2 passages were being considered, and a more extensive process of amalgamation can occur for some of the neighbouring passages. Additionally, also included in Table 5.7 is the Healing of the Paralytic passage which appeared in Table 5.6 and which passes the threshold of 80 words for Table 5.7 only after a process of amalgamation. It should also be noted that the Huck pericope number 212, Mk 12:41-44 || Lk 21:1-4, The Widow's Gift, that appears as part of one of the passages in Table 5.7, is a Mark-Luke double tradition pericope, absent from Matthew, so that it is not relevant to a comparison of the detailed use of Mark by Matthew and Luke.

5.4 Conclusions

Generally speaking, we would hope that the HMM approach to modelling would yield some insight into the way in which Matthew and Luke handled the text of Mark, especially through examination of the results of the decoding.

Table 5.7
Longer State 2 and State 4 passages for Luke's use of Mark (4 hidden states)

Passage in Mark	State	No. of words	Amalgamated and rounded	Description of passage
2:8:4-2:10:15	2	54	Mk 2:8-12	The Healing of
	3	12	(in Huck 52)	the Paralytic
2:11:10-2:13:2	4	28		
10:13:5-10:15:18	4	51	Mk 10:13-28	Jesus Blesses
10:16:1-10:17:11	2	20	(in Huck 188,	the Children,
	3	6	189)	The Rich
10:17:18-10:21:11	4	61		Young Man
	3	5		
10:21:17-10:23:8	2	35		
10:23:9-10:23:19	4	11		
10:23:20-10:29:3	2	87		
10:29:4-10:29:13	4	10		
12:17:6-12:24:4	4	115	Mk 12:18-23 (in Huck 207)	About the Resurrection
12:34:21-12:38:7	2	74	Mk 12:35-44	David's Son,
12:38:8-12:43:19	4	86	(in Huck 209,	Jesus Denounces
12:43:20-12:43:24	2	5	210,212)	Scribes, The
12:44:1-13:1:7	4	28		Widow's Gift
14:65:6-14:69:15	2	80	Mk 14:65-69 (in Huck 241)	Jesus Mocked, Peter's Denial

The work of this chapter represents a first attempt at fitting HMMs in the context of the synoptic problem, and there is doubtless scope for coming up with better HMMs in the future. For the models that we have fitted, at a simple level, for example, the longer State 1 passages identify sections of Mark which have been omitted by Matthew or Luke.

But we are principally concerned here with the issue of identifying passages where there appears to be strong evidence of dependence between Matthew and Luke in their verbal agreements with Mark. As a result of our statistical analysis and decoding, without any detailed textual analysis, we have arrived at a number of passages in Mark, those exhibited in Tables 5.4-5.7, which *prima facie* we may regard as suitable candidates for further investigation of the apparent positive association between Matthew's and Luke's verbal agreements with Mark, in order to assess the strength of the evidence for any claim that Matthew and Luke were not independent in their use of Mark. Because of the rather mechanical way, based on statistical modelling, in which these passages have been arrived at, we should not necessarily expect that all of them will provide useful insights into the synoptic problem, but we might hope that at least some of them will. We shall carry out an examination of a selection of these passages in Chapter 7 and will see that they do indeed turn out to be illuminating.

6

Mark's and Luke's use of Matthew

6.1 Introduction

In Chapters 3 and 5 we used as our base text the Gospel of Mark and, assuming Markan priority, analysed the binary time series that represented Matthew's and Luke's verbal agreements with Mark, especially with a view to investigating whether Matthew and Luke were independent in their use of Mark, as is supposed in the two-source hypothesis (see Figure 1.2).

In this chapter we carry out similar analyses but taking as our base text the Gospel of Matthew, so that the binary time series indicate which of the sequence of words in Matthew are also present unchanged in Mark and in Luke. According to the Griesbach hypothesis (see Figure 1.1) and according to the Augustinian hypothesis, both of which fall within the scope of the triple-link model, as described in Section 2.1, the Gospel of Matthew was the first of the synoptic gospels to be written and was used by both Mark and Luke. Under either of these two hypotheses, with their assumption of Matthean priority, the analyses of this chapter may be taken as an attempt to describe through the statistics of the verbal agreements how Mark and Luke used Matthew and how they may have been influenced by each other in their use of Matthew. Furthermore, according to the Farrer hypothesis (see Figure 1.3), the Gospel of Matthew was used by Luke, although not by Mark, so that we may at least regard our analyses of the binary series for Luke's verbal agreements with Matthew as attempting to give some insight into Luke's use of Matthew. However, even if we do not wish to work under the assumptions of any of these hypotheses, we may simply regard our analyses in this chapter as providing descriptions of the patterns of verbal agreements and non-agreements of Mark and Luke with Matthew. Even then, it is still of interest to explore the extent to which the models that are fitted here resemble the models for Mark as the base text, that were fitted in Chapters 3 and 5, and whether there are any readily apparent differences that are worth noting.

In Sections 6.2 to 6.5 of this chapter we shall present the results of analyses that parallel those of Chapters 3 and 5, but omitting some of the detail concerning the processes of model selection. In the final Section 6.6 we develop a new theme, showing how our data set with Matthew as the base text may be used to examine the so-called "minor agreements" of Matthew and Luke against Mark, which, under the assumption of Markan priority, are important in providing what may be taken to be evidence that Matthew and Luke were not independent in their use of Mark.

6.2 The data

In a way that parallels the construction of the data set described in Section 3.2 for Mark as the base text, the data set here is a word-by-word transcription of the colour-coded text of Matthew in Farmer's *Synopticon* into a bivariate binary time series that represents Mark's and Luke's verbal agreements with Matthew. The time series is of length 18300, which is the number of words in the Greek text of Matthew that was used by Farmer. The first component (x_t) of the bivariate time series is formed by writing $x_t = 1$ if the word in position t in the text of Matthew is present unchanged in Mark and $x_t = 0$ otherwise. The second component (y_t) is formed by writing $y_t = 1$ if the word is present unchanged in Luke and $y_t = 0$ otherwise. Thus the subscript t of the time series now refers to the position of the word in the text of Matthew.

Some remarks on the detail of the construction and checking of the data set are in order here. The data set, whether with Mark or Matthew as base text, was constructed by reading word by word through Farmer's text and translating the colour coding into the binary series. As a check, for each verse, the agreements from the binary series were aggregated to give verse totals of agreements, and these were compared with the verse totals of agreements tabulated in Chapters 2 and 3 of Tyson and Longstaff (1978) for Matthew and Mark as base text, respectively. Where there were discrepancies between the verse totals, a detailed check could be carried out. Discrepancies could occur for the various reasons listed in Section 1.4.2, but this check enabled some errors of the translation from colour coding into the binary series to be picked up. In the case of Mark as the base text, Farmer's colour coding was followed exactly, with the checking as described. In the light of the experience of working with this text and with warnings received about errors in Farmer's colour coding,[1] when it came to the construction of the data set with Matthew as the base text, a slightly more critical approach to Farmer's colour coding was taken. Where it seemed clear that there was a simple error in the colour coding, this was corrected, but in doubtful cases Farmer's coding was retained. The corrections were not many and are unlikely to affect the results of the analysis in any serious way.

The total numbers of ones and zeros represent overall counts of verbal agreements and non-agreements between Matthew and the other synoptic gospels. These counts are presented in Table 6.1 in the form of a contingency table, where in each cell, below the observed frequency, we have in brackets the

[1] In particular, from Mark Goodacre in private correspondence.

Table 6.1
Counts of verbal agreements with Matthew, and, in brackets, expected numbers under the hypothesis of statistical independence

		Luke		
		0	1	total
Mark	0	11225 (10387)	2363 (3201)	13588
	1	2764 (3602)	1948 (1110)	4712
	total	13989	4311	18300

expected frequency, to the nearest integer, under the hypothesis that Mark and Luke are statistically independent in their verbal agreements with Matthew.

By inspection of the row and column totals in Table 6.1 we see that, under the assumption of Matthean priority, overall Mark follows Matthew somewhat more closely than does Luke, retaining $4712/18300 = 26\%$ of Matthew's words unchanged, whereas Luke retains $4311/18300 = 24\%$ of Matthew's words unchanged. These proportions, 0.26 and 0.24, are also the mean values of the series (x_t) and (y_t), respectively.

The observed frequencies along the diagonal of Table 6.1 exceed the expected frequencies, which shows that Mark and Luke, if they are both using Matthew, make the same decision on whether to retain unchanged a word in Matthew more often than would be expected under the hypothesis of statistical independence. As discussed in Section 2.6, it is inappropriate to carry out a simple chi-square test, but the data in Table 6.1 do at least suggest that there may turn out to be significant evidence of a positive association between Mark and Luke in their verbal agreements with Matthew, and our time series analysis will confirm this.

Just as was done for Mark as the base text, we have constructed another binary time series (z_t) by writing $z_t = 1$ if the word in position t in the text of Matthew is part of the direct speech of Jesus or John or the divine voice and $z_t = 0$ if it is not part of such direct speech. This series will be used as an explanatory variable in our logistic regressions.

Using the pericopes as specified in Huck (1949), we partition the Gospel of Matthew into 167 pericopes that range in length from 14 to 597 words. To take into account that there may be variation in the way that Mark and Luke handle the different pericopes in Matthew, we shall later introduce a factor for pericope into our models.

For the present, to illustrate in outline the way in which the series (x_t) and (y_t) vary over the length of Matthew's gospel, in Figure 6.1 we provide a plot of the mean values of x_t (the solid line) and y_t (the dashed line) by pericope, where for the purposes of this plot the pericopes have been numbered 1 to 167 in the order in which they appear in Matthew's gospel. These means are, equivalently, the proportions of Matthew's words that are present unchanged in Mark and Luke, respectively. As may be seen from the plots, there are large fluctuations among the pericope means, even more so than in the corresponding Figure 3.2 for Matthew's and Luke's use of Mark.

The relatively small difference in the overall mean of 0.26 for (x_t) and 0.24 for (y_t), observed from the data of Table 6.1, masks a great deal of variation among the pericopes. For example, in the pericopes that make up the Matthew-Luke double tradition, the words in Matthew are altogether absent from Mark, so that the mean value of x_t is 0, but in the triple tradition pericopes Mark tends to have substantially more words in common with Matthew than does Luke, so that the mean of x_t tends to be much greater than the mean of y_t, and this is reflected in the plot of Figure 6.1, where the solid line tends to lie well above the dashed line for the triple tradition pericopes.

6.3 Logistic regression

Taking the approach of Sections 3.3 and 3.4, but now with Matthew as base text, we fit logistic regression models with similar predictor variables to the ones used previously. Consider firstly models for the series of verbal agreements of Mark with Matthew. Assuming that the series (x_t) has been observed up to the $(t-1)$th position, let π_t denote the probability that $X_t = 1$, i.e., that there is verbal agreement of Mark with Matthew in position t.

Let N_t^0 denote the length of the current run of 0s for the series (x_t) in position t and N_t^1 denote the length of the current run of 1s, where one or other of N_t^0 and N_t^1 will always be zero. As in Section 3.3, we use as predictor variables R_{t-1}^0 and R_{t-1}^1, where

$$R_t^0 = \ln(1 + N_t^0)$$

and

$$R_t^1 = \ln(1 + N_t^1) \ .$$

Additionally, we consider the lagged variables X_{t-1}, X_{t-2}, ... as potential predictor variables. Further, we may bring in the predictor variable Z_t for direct speech and the variable Y_t and other variables based on the series (y_t) for Luke's verbal agreements with Matthew such as S_{t-1}^0 and S_{t-1}^1, where S_t^0 and S_t^1 are the logarithms of the run lengths for the series (y_t), defined in exactly the same way as R_t^0 and R_t^1 for the series (x_t). Thus the model adopted is a logistic regression of the form

$$\ln\left(\frac{\pi_t}{1-\pi_t}\right) = \alpha + \beta_0 R_{t-1}^0 + \beta_1 R_{t-1}^1 + \dots, \qquad (6.1)$$

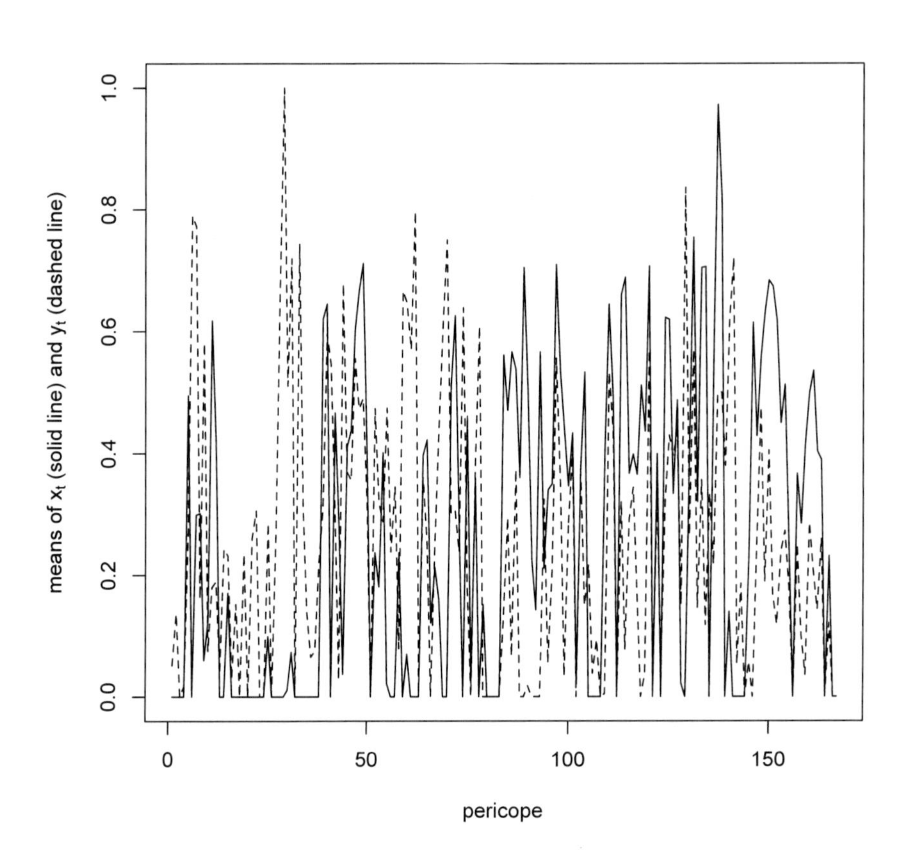

Figure 6.1
Plot of the mean values of x_t and y_t by pericope in Matthew

Table 6.2
Estimated regression coefficients for the series (x_t)

Model	Mk1	Mk2	Mk3	Mk4
constant	0.413	0.364	-0.261	-0.508
R^0_{t-1}	-1.103	-1.123	-1.187	-1.084
R^1_{t-1}	0.364	0.335	0.391	0.336
Z_t	–	0.192	0.111	0.204
Y_t	–	–	2.098	2.105
S^0_{t-1}	–	–	0.153	0.139
S^1_{t-1}	–	–	-0.450	-0.363
pericope factor	–	–	–	✓
residual deviance	11775	11758	10679	10510
residual d.f.	18295	18294	18291	(18289)

with a variety of possible predictor variables on the right-hand side of equation (6.1).

To begin with we consider as predictor variables the variables that are based on the previous history of the series (x_t): $R^0_{t-1}, R^1_{t-1}, X_{t-1}, \ldots$. A stepwise procedure starting with the null model gives as the single best predictor the variable R^0_{t-1}. At the next step the variable entered is R^1_{t-1}, but at the following step no significant improvement in fit is given by entering X_{t-1}. So we choose as our first fitted model the one with predictor variables R^0_{t-1} and R^1_{t-1}. The estimated regression coefficients for this model, Mk1, are shown in Table 6.2.

Next we include additionally the variable Z_t for direct speech as a predictor variable. Again following a stepwise procedure, the predictors $R^0_{t-1}, R^1_{t-1}, Z_t$ are successively entered, with the inclusion of Z_t giving a significant improvement in fit, but at the following step no significant improvement in fit is given by entering X_{t-1}. The estimated regression coefficients for the model Mk2 with predictor variables R^0_{t-1}, R^1_{t-1} and Z_t are shown in Table 6.2.

At the next stage we also include as predictors variables based on the series (y_t): Y_t, S^0_{t-1} and S^1_{t-1}. The stepwise procedure introduces the predictor variables in the order $R^0_{t-1}, Y_t, S^0_{t-1}, R^1_{t-1}, S^1_{t-1}, Z_t$, leading to the model Mk3 whose estimated regression coefficients are shown in Table 6.2. Comparison of the residual deviances for the models Mk3 and Mk2 shows that a very substantial improvement in fit has been obtained by introducing the predictor variables based on the series (y_t), so that there is a strong statistical dependence between the verbal agreements of Mark with Matthew and those of Luke with Matthew.

Finally, following the approach in Section 3.4, we introduce a normally distributed random effect $B_{H(t)}$ for pericope, where $H(t)$ denotes the pericope to which the word in position t belongs. In addition, because we are especially

Table 6.3
Estimated regression coefficients and standard errors for the model Mk4 for (x_t)

Variable	estimated coefficient	standard error	odds ratio
constant	-0.508	0.094	
R^0_{t-1}	-1.084	0.033	
R^1_{t-1}	0.336	0.043	
Y_t	2.105	0.129	8.208
S^0_{t-1}	0.139	0.025	
S^1_{t-1}	-0.363	0.058	
Z_t	0.204	0.065	1.227

interested in the dependence between (x_t) and (y_t) and how it might vary from pericope to pericope, we also introduce a normally distributed random interaction effect between Y_t and the pericope $H(t)$. The resulting generalized linear mixed model Mk4 has an estimated standard deviation of 0.486 for the main random effect, an estimated standard deviation of 1.016 for the interaction random effect, and estimated regression coefficients as given in Table 6.2. The decrease in residual deviance shows that the introduction of the random effects has led to a significant improvement in fit. The estimated regression coefficients for the model Mk4, together with their standard errors and the odds ratios for the binary regressor variables Y_t and Z_t, are also given in Table 6.3. The estimated coefficient 2.105 for Y_t is overwhelmingly significant as may be seen by comparing it with its standard error. So again we see that there is a strong statistical dependence, a positive association, between the verbal agreements of Mark with Matthew and those of Luke with Matthew.

We now consider, in a similar way, models for the series of verbal agreements of Luke with Matthew. Assuming that the series (y_t) has been observed up to the $(t-1)$th position, let θ_t denote the probability that $Y_t = 1$, i.e., that there is verbal agreement of Luke with Matthew in position t. We use as possible predictor variables the logarithms of the run lengths, S^0_{t-1} and S^1_{t-1}, the lagged variables $Y_{t-1}, Y_{t-2}, \ldots$, the variable Z_t for direct speech, and then the variable X_t and other variables based on the series (x_t) for Mark's verbal agreements with Matthew. Thus the model adopted is a logistic regression of the form

$$\ln\left(\frac{\theta_t}{1-\theta_t}\right) = \alpha + \beta_0 S^0_{t-1} + \beta_1 S^1_{t-1} + \ldots, \tag{6.2}$$

with a variety of possible predictor variables on the right-hand side of Equation (6.2).

To begin with we consider as predictor variables the variables that are

Table 6.4
Estimated regression coefficients for the series (y_t)

Model	Lk1	Lk2	Lk3	Lk4
constant	-0.195	-0.349	-1.359	-1.838
S^0_{t-1}	-0.893	-0.898	-0.908	-0.801
S^1_{t-1}	0.705	0.679	0.804	0.680
Z_t	–	0.290	0.325	0.314
X_t	–	–	2.053	2.264
R^0_{t-1}	–	–	0.170	0.196
R^1_{t-1}	–	–	-0.382	-0.202
pericope factor	–	–	–	✓
residual deviance	12801	12761	11713	11432
residual d.f.	18295	18294	18291	(18289)

based on the previous history of the series (y_t): $S^0_{t-1}, S^1_{t-1}, Y_{t-1}, \ldots$. A stepwise procedure starting with the null model gives as the single best predictor the variable S^0_{t-1}. At the next step the variable entered is S^1_{t-1}, but at the following step no significant improvement in fit is given by entering Y_{t-1}. So we choose as our first fitted model the one with predictor variables S^0_{t-1} and S^1_{t-1}. The estimated regression coefficients for this model, Lk1, are shown in Table 6.4.

Next we include additionally the variable Z_t for direct speech as a predictor variable. Again following a stepwise procedure, the predictors $S^0_{t-1}, S^1_{t-1}, Z_t$ are successively entered, with the inclusion of Z_t giving a significant improvement in fit, but at the following step no significant improvement in fit is given by entering Y_{t-1}. The estimated regression coefficients for the model Lk2 with predictor variables S^0_{t-1}, S^1_{t-1} and Z_t are shown in Table 6.4.

At the next stage we also include as predictors variables based on the series (x_t): X_t, R^0_{t-1} and R^1_{t-1}. The stepwise procedure introduces the predictor variables in the order $S^0_{t-1}, X_t, S^1_{t-1}, R^0_{t-1}, R^1_{t-1}, Z_t$, leading to the model Lk3 whose estimated regression coefficients are shown in Table 6.4. Comparison of the residual deviances for the models Lk3 and Lk2 shows that a very substantial improvement in fit has been obtained by introducing the predictor variables based on the series (x_t), so we see again that there is a strong statistical dependence between the verbal agreements of Luke with Matthew and those of Mark with Matthew.

Finally, as in the modelling of the verbal agreements of Mark with Matthew, we introduce the normally distributed random effect $B_{H(t)}$ for pericope and a normally distributed random interaction effect between X_t and the pericope $H(t)$. The resulting generalized linear mixed model Lk4 has an estimated standard deviation of 0.820 for the main random effect, an estimated standard deviation of 1.044 for the interaction random effect, and estimated regression coefficients as given in Table 6.4. The decrease in residual deviance

Table 6.5
Estimated regression coefficients and standard errors for the model Lk4 for (y_t)

Variable	estimated coefficient	standard error	odds ratio
constant	-1.838	0.123	
S^0_{t-1}	-0.801	0.029	
S^1_{t-1}	0.680	0.044	
X_t	2.264	0.125	9.621
R^0_{t-1}	0.196	0.027	
R^1_{t-1}	-0.202	0.047	
Z_t	0.314	0.069	1.369

shows that the introduction of the random effects has led to a significant improvement in fit. The estimated regression coefficients for the model Lk4, together with their standard errors and the odds ratios for the binary regressor variables X_t and Z_t are also given in Table 6.5. The estimated coefficient 2.264 for X_t is overwhelmingly significant as may be seen by comparing it with its standard error. So yet again we see a strong positive association between the verbal agreements of Luke with Matthew and those of Mark with Matthew.

In summary, the models that we have fitted and presented in Tables 6.2-6.5 with Matthew as the base text are broadly in line with those of Sections 3.3 and 3.4 for Mark as the base text. No immediately obvious new features arise. However, just as in Sections 3.3 and 3.4, a by-product of the analysis of the generalized linear mixed models, here Mk4 and Lk4, is that we can examine the interactions with the pericope factor of the predictors Y_t and X_t, respectively. Figure 6.2 gives a scattergram of the predicted interactions for the individual pericopes. Those pericopes for which these interactions are largest in a positive direction are the ones where the dependence between the series (x_t) and (y_t) appears to be the strongest. The two pericopes that stand out in the top right-hand corner of Figure 6.2 are presented in Table 6.6. These pericopes, different from the ones presented in Table 3.8 in Section 3.4, are suggested by our analysis as ones that may be of interest for more detailed examination to see how it is that there is in them a particularly strong positive association between Mark and Luke in their patterns of verbal agreement with Matthew.

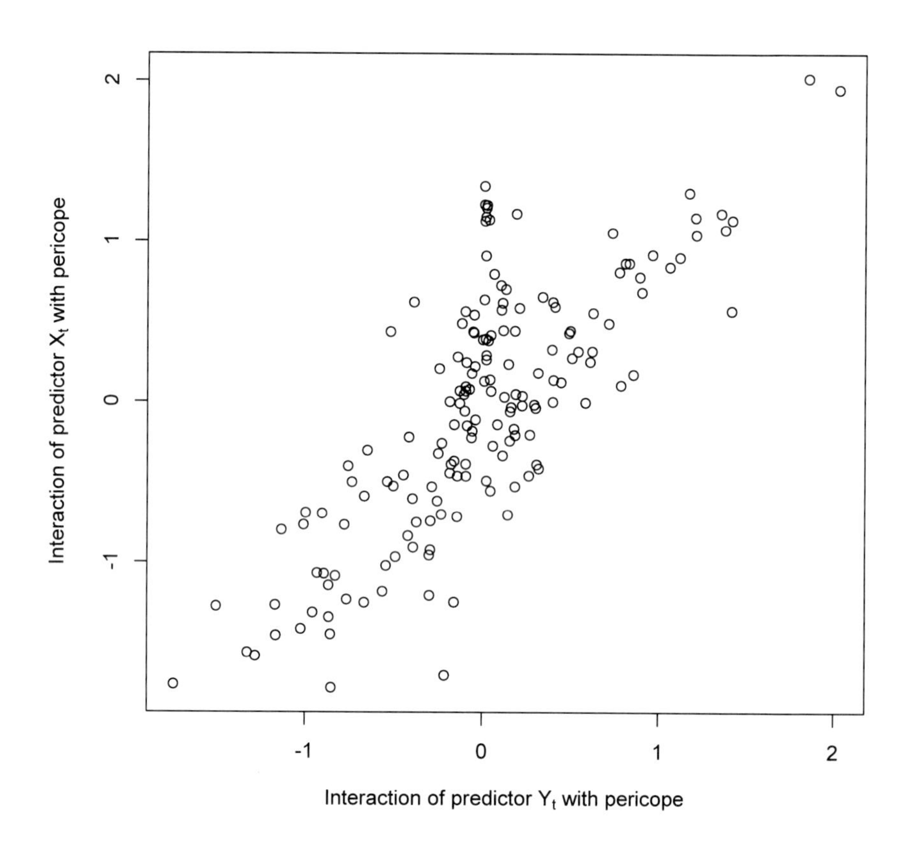

Figure 6.2
Scatter plot of interactions of the predictors X_t, Y_t with the pericope effect

Table 6.6
Pericopes with large positive interactions

Huck number	Pericope description	Passage in Matthew	Interactions: with Y_t in Mk4	with X_t in Lk4
51	The Gadarene Demoniacs	8:28-34	2.024	1.946
193	Healing of Bartimaeus	20:29-34	1.851	2.017

6.4 Hidden Markov models

We now turn to the use of hidden Markov models with Matthew as the base text. Assuming for the present the priority of Matthew, it is natural to interpret the hidden states as representing different modes of behaviour by Mark and Luke in their use of Matthew. We use the notation and theory developed in Chapter 4 in a similar way to its use in Chapter 5, where Mark was the base text, but in the present section we include in our presentation rather less detail than was the case in Chapter 5.

We examine first the series (x_t) for Mark's use of Matthew. Because there are substantial sections of Matthew, Matthew's single tradition and the Matthew-Luke double tradition, where Mark is not using Matthew at all, it is natural to take one hidden state, State 1, to be such that Mark is not using Matthew, so that $\pi_t^1 = 0$. To obtain useful models we then need at least two further hidden states to represent different modes of behaviour in Mark's use of Matthew, so that we have at least three hidden states in total. The use of four hidden states is also explored, although then, because of the number of parameters involved, problems of convergence of the algorithm for maximum likelihood estimation begin to emerge. Useful regressor variables turn out to be the lagged variable X_{t-1}, the variable Z_t for direct speech and the variable Y_t for Luke's use of Matthew. As we noted in Chapter 5, the role of the "log-run length" variables, R_{t-1}^0 and R_{t-1}^1, so significant in the logistic regression modelling of Chapter 3 and Section 6.3, is essentially taken over by the hidden states.

In Table 6.7 we summarise the results of fitting these models for three and four hidden states, $m = 3$ and $m = 4$, respectively, and for all subsets of the regressor variables X_{t-1}, Z_t, Y_t, where k denotes the number of regressor variables used. We see that the model that gives the minimum BIC is the one with three hidden states and all three regressor variables, but the model that gives the minimum AIC is the one with four hidden states and all three regressor variables. In both cases we have a relatively large value of the mean stay μ, which suggests that these models may, after decoding, give us reasonably long pieces of text with the same hidden state, as discussed in Section 4.5.

In Equations (6.3)-(6.6) we give the fitted model for $m = 3$ and the regressor variables X_{t-1}, Z_t, Y_t, the model preferred according to the BIC criterion. The fitted transition matrix $\mathbf{\Gamma}$ for the hidden states is given by

$$\mathbf{\Gamma} = \begin{pmatrix} 0.9912 & 0.0065 & 0.0023 \\ 0.0177 & 0.8668 & 0.1154 \\ 0.0037 & 0.0694 & 0.9269 \end{pmatrix}. \tag{6.3}$$

Table 6.7
Comparison of fitted HMMs for Mark's use of Matthew, (X_t)

m	k	free/working parameters	regressor variables	$-\ln L_T$	AIC	BIC	mean stay μ	model chosen
3	0	8		5902.48	11820.96	11883.48	10.12	
3	1	10	Y_t	5355.48	10730.96	10809.11	11.36	
3	1	10	X_{t-1}	5896.43	11812.87	11891.01	16.35	
3	1	10	Z_t	5860.67	11741.34	11819.48	10.03	
3	2	12	X_{t-1}, Z_t	5851.54	11727.08	11820.85	15.03	
3	2	12	Z_t, Y_t	5308.46	10640.92	10734.70	11.40	
3	2	12	X_{t-1}, Y_t	5290.43	10604.86	10698.63	20.43	
3	3	14	X_{t-1}, Z_t, Y_t	5253.21	10534.41	**10643.81**	19.31	✓
4	0	15		5871.90	11773.79	11891.01	9.33	
4	1	18	Y_t	5267.46	10570.92	10711.59	9.82	
4	1	18	X_{t-1}	5863.22	11762.44	11903.10	27.94	
4	1	18	Z_t	5825.11	11686.23	11826.89	9.95	
4	2	21	X_{t-1}, Z_t	5798.74	11639.48	11803.58	13.78	
4	2	21	Z_t, Y_t	5224.77	10491.54	10655.65	9.85	
4	2	21	X_{t-1}, Y_t	5247.46	10536.91	10701.02	24.96	
4	3	24	X_{t-1}, Z_t, Y_t	5213.43	**10474.86**	10662.41	20.55	✓

The fitted model equations are

$$\pi_t^1 = 0 \tag{6.4}$$

$$\ln\left(\frac{\pi_t^2}{1-\pi_t^2}\right) = -3.577 + 0.834X_{t-1} - 0.307Z_t + 6.577Y_t \tag{6.5}$$

$$\ln\left(\frac{\pi_t^3}{1-\pi_t^3}\right) = -0.385 + 1.154X_{t-1} + 0.694Z_t + 0.495Y_t \tag{6.6}$$

The signs of the coefficients on the right-hand sides of Equations (6.5) and (6.6) show that, under the assumption of Matthean priority, in both State 2 and State 3 a word in Matthew is more likely to be retained unchanged by Mark if the previous word was retained by Mark and if the current word was retained by Luke. Comparing the sizes of the coefficients, we observe that State 2 appears to represent an overall less intensive use of Matthew by Mark than does State 3. However, the large coefficient 6.577 for Y_t in Equation (6.5) shows that, in State 2, if a word in Matthew is retained unchanged by Luke then it has a high probability of being retained unchanged by Mark, so that in State 2 there is a particularly strong positive association between whether a word in Matthew is retained by Mark and whether it is retained by Luke. So it may be anticipated that, after decoding, the State 2 passages might provide examples of texts with substantial evidence of dependence between Mark and Luke in their use of Matthew. It may also be noted that the sign of the coefficient of Z_t, although positive in State 3, is negative in State 2, so that in State 2 there is some negative association between direct speech and retention by Mark.

In Equations (6.7)-(6.11) we give the fitted model for $m = 4$ and the regressor variables X_{t-1}, Z_t, Y_t, the model preferred according to the AIC criterion. The fitted transition matrix $\mathbf{\Gamma}$ for the hidden states is given by

$$\mathbf{\Gamma} = \begin{pmatrix} 0.9945 & 0.0038 & 0.0016 & 0.0000 \\ 0.0000 & 0.9465 & 0.0375 & 0.0160 \\ 0.0124 & 0.0213 & 0.8674 & 0.0989 \\ 0.0012 & 0.0033 & 0.0599 & 0.9356 \end{pmatrix}. \tag{6.7}$$

The fitted model equations are

$$\pi_t^1 = 0 \tag{6.8}$$

$$\ln\left(\frac{\pi_t^2}{1-\pi_t^2}\right) = -1.487 + 1.290X_{t-1} - 27.378Z_t + 1.116Y_t \tag{6.9}$$

$$\ln\left(\frac{\pi_t^3}{1-\pi_t^3}\right) = -4.629 + 0.220X_{t-1} + 0.805Z_t + 7.581Y_t \tag{6.10}$$

$$\ln\left(\frac{\pi_t^4}{1-\pi_t^4}\right) = -0.284 + 1.101X_{t-1} + 0.591Z_t + 0.694Y_t \tag{6.11}$$

The signs of the coefficients on the right-hand sides of Equations (6.9)-(6.11) show that, under the assumption of Matthean priority, in State 2, State 3 and State 4 a word in Matthew is more likely to be retained unchanged by Mark if the previous word was retained by Mark and if the current word was retained by Luke. Comparing the sizes of the coefficients, especially of the constant terms, State 3 appears to represent an overall less intensive use of Matthew by Mark than do States 2 and 4, but with the important qualification that the large coefficient 7.581 for Y_t in Equation (6.10) shows that in State 3 there is a particularly strong positive association between whether a word in Matthew is retained by Mark and whether it is retained by Luke. So it may be anticipated that, after decoding, the State 3 passages might provide examples of texts with substantial evidence of dependence between Mark's and Luke's use of Matthew. The negative coefficient for Z_t in Equation (6.9) is so large that in State 2, when $Z_t = 1$, π_t is microscopically small, so that the direct speech in Matthew is simply not retained by Mark. It turns out that some of the decoded State 2 passages do include pieces of direct speech in Matthew, which are not present in Mark. Nevertheless, the existence of such an exceedingly large coefficient, whose presence overwhelms everything else in the model equation for State 2 when $Z_t = 1$, is an unappealing feature of this particular fitted model. It also leads to problems of convergence of the maximum likelihood algorithm for model fitting.

To summarise these conclusions regarding our chosen HMM models for Mark's verbal agreements with Matthew in a way that is not dependent upon the assumption of Matthean priority, for the three-state model it is in the State 2 passages and for the four-state model it is in the State 3 passages that we might expect to find the strongest similarities between Luke and Mark in the patterns of their verbal agreements and non-agreements with Matthew.

For Luke's use of Matthew, we examine the series (y_t). In Table 6.8 we summarise the results of fitting HMMs, again for three and four hidden states but now for all subsets of the regressor variables Y_{t-1}, Z_t, X_t. There are substantial passages in Matthew that are not used by Luke, and, just as for Mark's use of Matthew, it turns out to be appropriate to reserve State 1 for passages where Luke is not using Matthew at all, so that $\theta_t^1 = 0$. In a way that parallels the models that we chose for Mark's use of Matthew, we choose the models with all the regressor variables Y_{t-1}, Z_t, X_t ($k = 3$) and with either three hidden states ($m = 3$) or four hidden states ($m = 4$). But here, for Luke's use of Matthew, whether by the AIC or BIC criterion, the model with $m = 4$ is preferred, although the model for $m = 3$ has a larger value for the mean stay.

In Equations (6.12)-(6.15) we present the fitted model for the case $m = 3$. The fitted transition matrix $\boldsymbol{\Delta}$ for the hidden states is given by

$$\boldsymbol{\Delta} = \begin{pmatrix} 0.9774 & 0.0218 & 0.0007 \\ 0.0225 & 0.9369 & 0.0406 \\ 0.0067 & 0.0598 & 0.9336 \end{pmatrix}. \tag{6.12}$$

Table 6.8
Comparison of fitted HMMs for Luke's use of Matthew, (X_t)

m	k	free/working parameters	regressor variables	$-\ln L_T$	AIC	BIC	mean stay μ	model chosen
3	0	8		6456.83	12929.66	12992.17	10.22	
3	1	10	X_t	6004.59	12029.18	12107.33	12.87	
3	1	10	Y_{t-1}	6406.69	12833.39	12911.53	29.49	
3	1	10	Z_t	6425.06	12870.12	12948.26	9.98	
3	2	12	Y_{t-1}, Z_t	6373.16	12770.33	12864.10	25.63	
3	2	12	Z_t, X_t	5966.07	11956.13	12049.90	11.54	
3	2	12	Y_{t-1}, X_t	5853.98	11731.96	11825.74	23.06	
3	3	14	Y_{t-1}, Z_t, X_t	5821.46	11670.92	11780.32	21.38	✓
4	0	15		6404.44	12838.88	12956.10	8.29	
4	1	18	X_t	5846.59	11729.17	11869.83	9.28	
4	1	18	Y_{t-1}	6361.41	12758.82	12899.48	12.98	
4	1	18	Z_t	6375.09	12786.17	12926.84	7.81	
4	2	21	Y_{t-1}, Z_t	6329.77	12701.55	12865.65	14.24	
4	2	21	Z_t, X_t	5822.11	11686.21	11850.32	9.18	
4	2	21	Y_{t-1}, X_t	5764.30	11570.60	11734.71	16.21	
4	3	24	Y_{t-1}, Z_t, X_t	5738.91	**11525.83**	**11713.38**	15.45	✓

The fitted model equations are

$$\theta_t^1 = 0 \tag{6.13}$$

$$\ln\left(\frac{\theta_t^2}{1-\theta_t^2}\right) = -2.912 + 1.128Y_{t-1} + 0.398Z_t + 3.484X_t \tag{6.14}$$

$$\ln\left(\frac{\theta_t^3}{1-\theta_t^3}\right) = -1.049 + 1.867Y_{t-1} + 0.873Z_t - 0.896X_t \tag{6.15}$$

The signs of the coefficients on the right-hand sides of Equations (6.14) and (6.15) show that in both State 2 and State 3 a word in Matthew is more likely to be retained unchanged by Luke if the previous word was retained by Luke and if the current word is part of the direct speech of Jesus. State 2 appears to represent a generally less intensive use of Matthew by Luke than does State 3, but the large positive coefficient 3.484 for X_t in Equation (6.14) shows that in State 2 there is a strong positive association between whether a word in Matthew is retained by Luke and whether it is retained by Mark. In Equation (6.15) the negative coefficient for X_t shows that in State 3 for Luke's use of Matthew, a word is less likely to be retained by Luke if it is retained by Mark, so that there is some negative association between Mark's and Luke's retentions, which is similar to what was seen in Equation (5.14) for the three-state HMM for Luke's use of Mark.

In Equations (6.16)-(6.20) we present the fitted model for the case $m = 4$. The fitted transition matrix $\boldsymbol{\Delta}$ for the hidden states is given by

$$\boldsymbol{\Delta} = \begin{pmatrix} 0.9786 & 0.0178 & 0.0032 & 0.0004 \\ 0.0219 & 0.8882 & 0.0858 & 0.0042 \\ 0.0095 & 0.0801 & 0.9016 & 0.0088 \\ 0.0073 & 0.0000 & 0.0275 & 0.9652 \end{pmatrix}. \tag{6.16}$$

The fitted model equations are

$$\theta_t^1 = 0 \tag{6.17}$$

$$\ln\left(\frac{\theta_t^2}{1-\theta_t^2}\right) = -4.305 + 0.989Y_{t-1} + 0.289Z_t + 5.663X_t \tag{6.18}$$

$$\ln\left(\frac{\theta_t^3}{1-\theta_t^3}\right) = -1.676 + 1.414Y_{t-1} + 0.655Z_t + 0.165X_t \tag{6.19}$$

$$\ln\left(\frac{\theta_t^4}{1-\theta_t^4}\right) = -0.558 + 1.782Y_{t-1} + 0.898Z_t + 2.102X_t \tag{6.20}$$

In this case, the fitted coefficients of all three regressor variables, for each of the States 2, 3 and 4, are positive, so that there is a positive association between whether a word is retained unchanged by Luke and whether it is part of the direct speech, and whether it has been retained by Mark, and whether the previous word has been retained unchanged by Luke. Generally in State 2 Luke follows Matthew less closely than in States 3 and 4, but the relatively large regression coefficient 5.663 for X_t shows that in State 2 the positive association between whether a word is retained by Luke and whether it is retained by Mark is particularly strong.

So, under the assumption of Matthean priority, whether we use the HMM with three hidden states or with four hidden states for Luke's use of Matthew, after decoding it is in the State 2 passages that we might expect to find the strongest evidence of dependence between Luke and Mark in their use of Matthew. To put it in a way that is not dependent upon the assumption of Matthean priority, it is in the State 2 passages that we might expect to find the strongest similarities between Luke and Mark in the patterns of their verbal agreements and non-agreements with Matthew and, perhaps, substantial agreements of Luke and Mark against Matthew.

6.5 Decoding

We begin with the fitted HMM with three hidden states for Mark's use of Matthew, as described in Equations (6.3)-(6.6), and focus on the State 2 passages, where the association between Mark's and Luke's use of Matthew may be expected to be at its strongest. When the global decoding is carried out, there are 174 State 2 passages with a total of 2946 words, varying in length from 1 to 91 words and with a median of 13.0. If we now consider which of these passages might merit closer examination, as in Section 5.3 we reject the shorter passages as being unlikely to provide sufficient material for any useful discussion, but we amalgamate longer State 2 passages with neighbouring State 2 passages which are separated from them by only a few words with a different hidden state or states. In Table 6.9 there are presented the State 2 passages for Mark's use of Matthew which are at least 70 words in length. Shorter State 2 passages that are near neighbours are also included and merged. This results in a total of 4 passages for consideration, which are then rounded so that only whole verses are included in a passage. The first column in Table 6.9 gives a precise specification of the State 2 passages. The second column gives the exact number of words in each passage before amalgamation and rounding, and also includes the number of words in intervening passages. The third column gives the specification of the amalgamated and rounded passage, which appears in its own box in the table together with the number of the Huck pericope in which it is contained, and the fourth column gives a description of the passage.

When global decoding is carried out for the fitted HMM with four hidden states for Mark's use of Matthew, as specified in Equations (6.7)-(6.11), it is

Table 6.9
Longer State 2 passages for Mark's use of Matthew (3 hidden states)

Passage in Matthew	No. of words	Amalgamated and rounded	Description of passage
8:28:7-8:28:16	10	Mt 8:28-34	The Gadarene
	2	(Huck 51,	Demoniacs
8:28:19-8:32:21	75	cf Huck 106)	
	9		
8:33:3-8:34:17	30		
12:4:2-12:10:3	91	Mt 12:4-8 (in Huck 69)	Plucking Heads of Grain on the Sabbath
13:12:15-13:15:30	70	Mt 13:13-15 (in Huck 91)	The Reason for Speaking in Parables
19:16:12-19:21:8	74	Mt 19:16-21 (in Huck 189)	The Rich Young Man

the State 3 passages where the association between Mark's and Luke's use of Matthew may be expected to be at its strongest. There are 148 State 3 passages with a total of 2534 words, varying in length from 1 to 74 words and with a median of 13.0. The longer State 3 passages are very similar to the State 2 passages for the three-state model, and we do not tabulate them here.

Turning now to the models for Luke's use of Matthew, when the global decoding is carried out for the fitted HMM with three hidden states as described in Equations (6.12)-(6.15), it is the State 2 passages where the association between Luke's and Mark's use of Matthew may be expected to be at its strongest. There are 188 State 2 passages with a total of 6818 words, varying in length from 1 to 350 words and with a median of 22, but 14 of these passages exceed 100 words in length.

We now consider the fitted HMM with four hidden states for Luke's use of Matthew, as described in Equations (6.16)-(6.20), and again focus on the State 2 passages. Global decoding gives 158 State 2 passages for Luke's use of Matthew with a total of 3989 words, varying in length from 1 to 150 words and with a median of 17. There are 5 passages that exceed 100 words in length, a more manageable number of passages than the 14 that arose from the model with three hidden states. Two of these 5 are passages where Mark does not use Matthew at all, so they are not going to be relevant for discussing the relationship of Luke and Mark in their detailed use of Matthew. This leaves 3 passages, but we include the next longest State 2 passage, the Healing of Bartimaeus, which was identified as a passage of interest in Table 6.6. In this case there are no neighbouring State 2 passages to be amalgamated, but some rounding will be in order to include whole verses. In Table 6.10 the first column gives a precise specification of the State 2 passages. The second column gives

Table 6.10
Longer State 2 passages for Luke's use of Matthew (4 hidden states)

Passage in Matthew	No. of words	Rounded	Description of passage
8:28:7-8:34:17	126	Mt 8:28-34 (Huck 51, cf Huck 106)	The Gadarene Demoniacs
19:13:3-19:21:20	140	Mt 19:13-21 (in Huck 188, 189)	Jesus Blesses Children, The Rich Young Man
20:29:5-21:1:16	91	Mt 20:29-34 (Huck 193)	Healing of Bartimaeus
23:1:10-23:12:3	150	Mt 23:1-12 (in Huck 210)	Jesus Denounces Scribes and Pharisees

the exact number of words in each passage before rounding. The third column gives the specification of the rounded passage, which appears in its own box in the table together with the number or numbers of the Huck pericopes in which it is contained, and the fourth column gives a description of the passage.

As a result of our statistical analysis, without any detailed textual analysis, we have thus arrived at a number of passages in Matthew, those exhibited in Tables 6.6, 6.9 and 6.10, which, under the assumption of Matthean priority, we may *prima facie* regard as suitable candidates for further investigation of the apparent positive association between Mark's and Luke's use of Matthew. To put it more neutrally, without any assumption of which gospel came first, we may examine these passages, where we might expect to find a particularly strong association between the patterns of verbal agreements and non-agreements of Mark with Matthew and of Luke with Matthew, in order to see if they might give us any insight into aspects of the synoptic problem. We shall carry out an examination of some of them in Chapter 7.

6.6 The minor agreements

In the triple tradition of pericopes that are present in all three synoptic gospels, what are known as the minor agreements of Matthew and Luke against Mark, or just the "minor agreements", are the instances where Matthew and Luke, in some more or less precisely defined aspect, agree in the detail of their wording, but differ from Mark. These minor agreements have long been a source of embarrassment for supporters of the two-source hypothesis as they may be argued to provide evidence that Matthew and Luke did not work independently in their use of Mark.

The minor agreements have been studied by many authors. For example, a list is to be found in Hawkins (1909), pp. 208-212, and the issue is discussed in Chapter 11 of Streeter (1924), but a comprehensive list of various types of minor agreement, drawing on the work of earlier authors, is provided by Neirynck (1974) and, in a simplified version with textual notes and other material omitted, in Neirynck (1991). The classification of the minor agreements into different types and methods for counting them have been reviewed and discussed in Vinson (2004).

The main thrust of the present study has not involved consideration of the minor agreements. Nevertheless they do provide important supplementary input and have provoked much discussion in the biblical studies literature, as we shall see when we come to examine particular pieces of text in Chapter 7. It is certainly worthwhile to comment on how the data sets that we have created relate to the issue of the minor agreements. Our data set with Mark as the base text, that was used in our analyses in Chapters 3 and 5, could only provide us with very limited information on these minor agreements, because there was no direct comparison between the texts of Matthew and Luke. The only kind of agreement that could be observed was where Matthew and Luke agreed in not retaining unchanged a word present in Mark. However, the data set with Matthew as base text can take us much further in that it records those words that Matthew and Luke have in common but that Mark does not have. In terms of the notation of this chapter, we consider the words in the set of positions t in the Gospel of Matthew that are specified as

$$\{t : x_t = 0 \ \& \ y_t = 1\},$$

which are the words in Matthew that are absent from Mark but present in Luke. If in addition we restrict attention to the triple tradition of pericopes, we obtain a list of exact minor agreements in the wording of Matthew and Luke against Mark in the very precise sense of words that are present in exactly the same grammatical form in Matthew and Luke but not in Mark. In the terminology of Neirynck (1974), p. 52, these agreements fall within the category of “agreements in addition or substitution (positive agreements)”. In the terminology of Vinson (2004), they are either “common insertions” or “common changes”.

On carrying out this procedure, we found that there were 747 such verbal agreements of Matthew and Luke against Mark. In Table 6.11 we list the three pericopes with the largest proportions of such agreements, i.e., the largest values of the ratio shown in the final column of the table, the ratio of the total number of agreements to the total number of words in Matthew's version of the pericope. Examination of these pericopes reveals that we have to be a little more careful in our specification of what we may refer to as the “minor verbal agreements”, because these pericopes are all of a kind that appear to be a mixture of triple tradition and double tradition material — they all contain

Table 6.11
Major and minor verbal agreements of Matthew and Luke against Mark

Huck no.	Pericope description	Passage in Mt	Words in Mt	Agreements total	Agreements minor	Ratio
4	John's Preaching	3:11-12	57	29	4	51%
8	The Temptation	4:1-11	184	103	3	56%
86	A House Divided	12:25-37	262	105	7	40%

substantial segments of material that are present in Matthew and Luke but absent from Mark. As a result there are in these segments large numbers of verbal agreements that we may refer to as "major agreements" of Matthew and Luke against Mark, as discussed by Goodacre (2001), pp. 52-53. The pericopes listed in our Table 6.11 are among those listed in Goodacre's Table 7. For proponents of the two-source hypothesis, these are pericopes where there is a Mark-Q overlap, where Matthew and Luke have drawn on both Mark and Q in their versions of the pericope.

As a method of filtering out these major agreements, we have used the results of our decoding of the fitted HMM with three hidden states for Mark's use of Matthew. The passages in Matthew that correspond to State 1 are ones where there are no verbal agreements between Mark and Matthew and consist mainly of passages where there is no parallel passage in Mark. We may expect that elimination of these passages from our counts of agreements will remove all, or at least nearly all, of what we may regard as the major agreements of Matthew and Luke against Mark. What verbal agreements remain after removal of these State 1 passages, we refer to as minor verbal agreements. For the pericopes in Table 6.11, we see that there are very few such minor verbal agreements. Overall, the 747 verbal agreements of Matthew and Luke against Mark are reduced to 370 minor verbal agreements.

As a further step in this procedure we eliminate those pericopes in Matthew where there is a discrepancy between the total number of verbal agreements and the minor verbal agreements. What then remain are what we may regard as purely triple tradition pericopes, where all the verbal agreements of Matthew and Luke against Mark may be taken to be genuinely minor agreements that, under the assumption of Markan priority, reflect the detailed handling of the text of Mark by Matthew and Luke. After this step has been taken, there remain 240 minor verbal agreements, and the pericopes with the largest proportions of such agreements are listed in Table 6.12. From the viewpoint of the minor agreements, these pericopes are ones that are signalled as potentially being of special interest for investigation of the question of the independence of Matthew and Luke in their use of Mark.

Table 6.12
Minor verbal agreements of Matthew and Luke against Mark

Huck no.	Pericope description	Passage in Mt	Words in Mt	Agree-ments	Ratio
52	The Healing of the Paralytic	9:1-8	126	13	10%
112	The Feeding of the Five Thousand	14:13-21	158	15	9%
208	The Great Commandment	22:34-40	81	14	17%
241	Jesus Before the Council	26:57-75	312	31	10%
242	Jesus Delivered to Pilate	27:1-2	28	4	14%

7

Examples of synoptic parallels

7.1 Introduction

In Chapters 3 and 5 we worked with binary time series created from the texts of the synoptic gospels. As a result of our statistical analyses, under the assumption of Markan priority, we arrived at tables that suggested pericopes and more specific sections of text that might be of interest in investigating the statistical dependence of Matthew and Luke in their use of Mark: Table 3.8 that arose from the logistic regression approach and Tables 5.4, 5.5, 5.6 and 5.7 that arose from the decoding following on from the fitting of hidden Markov models. Furthermore, in Chapter 6, Table 6.12 suggested pericopes for investigation based on the frequencies of minor agreements, in the sense of exact verbal agreements of Matthew and Luke against Mark.

We are now ready to put back the flesh of the biblical texts on the dry bones of the statistical data and turn to the detailed examination of some of the individual sections of text that are indicated in these tables. In so doing, we shall engage with some examples of the arguments used by New Testament scholars to support the synoptic hypotheses that they favour, in interplay with the results of our statistical analysis. It is important too to see how useful any apparent insights gained from our analysis of the statistical models, with their simplifying assumptions, turn out to be when faced with the complexities of the texts themselves. The central issue to be dealt with here is whether, under the assumption of Markan priority, given the statistical dependence of Matthew and Luke in their use of Mark, as established in Chapter 3, it is also the case that Luke was dependent on Matthew or vice versa, in the sense of one of them having the other as a source. As we shall see when faced in particular cases with the arguments for and against such dependence, this is a much more subtle issue than that of statistical dependence, and not amenable to any form of statistical analysis that could lead to conclusions framed in the quantitative terms of numerical probabilities or significance levels. Our conclusions will necessarily be qualitative rather than quantitative. Although our statistical data will still have a role to play, there are limits to how far they will take us.

To limit the number of pericopes to be examined in detail to what is a manageable number in the context of the present monograph, in Table 7.1 we list the pericopes that appear in at least two of the above-mentioned tables,

Table 7.1
Candidate pericopes for investigating Matthew's and Luke's use of Mark

Huck no.	Pericope description	Tables in which present
45	The Healing of a Leper	3.8, 5.4/5.5
52	The Healing of the Paralytic	5.6/5.7, 6.12
112	The Feeding of the Five Thousand	5.4/5.5, 6.12
208	The Great Commandment	3.8, 6.12
209	About David's Son	3.8, 5.6/5.7
241	Jesus Before the Council	5.7, 6.12

but counting a double occurrence in Tables 5.4/5.5 for Matthew's use of Mark only once, as all the passages in Table 5.4 appear also in the related Table 5.5. Similarly we count a double occurrence in Tables 5.6/5.7 for Luke's use of Mark only once, as both the passages in Table 5.6 appear also in Table 5.7. The detailed examination of the passages in Table 7.1 is carried out in the following sections of the present chapter. In each case, for a particular section of text, the parallel Greek versions according to the three synoptic gospels are presented in one figure, using the NA^{25} critical text,[1] and the parallel English versions according to the NRSV translation in another figure. For both the Greek texts and the English texts, words that are present in all three synoptic gospels are double-underlined, words that are present in Mark and Matthew alone are single-underlined, and words that are present in Mark and Luke alone have a wavy single underline, as shown in Figure 7.1. In addition, the minor verbal agreements of Matthew and Luke against Mark, of words that are present in Matthew and Luke, but absent from Mark, are enclosed in boxes. The underlinings for the Greek texts match exactly the verbal agreements as recorded in the time series data sets that we have constructed, whereas the underlinings for the English translations will often give only a very approximate idea of the agreements in the Greek text.

For general background and input to the examination and comparison of the texts it is useful to consult commentaries, of which there are a number of suitable ones available, often written from the standpoint of differing models of synoptic relationships. We have made particular use of the following: for Mark, Hooker (1991) and France (2002); for Matthew, Gundry (1994) and Davies and Allison (1997); for Luke, Marshall (1978) and Goulder

[1]As noted in Section 1.4.2, Farmer (1969) used the NA^{25} text in constructing his *Synopticon*, and, as described in Sections 3.2 and 6.2, we constructed our binary time series data sets from the colour coding of the *Synopticon*. In order that the texts that we consider in this chapter correspond precisely to the statistical data that we have analysed, we have used the NA^{25} text here too. Differences from the NA^{28} occur only occasionally, and where any difference is of significance for the synoptic problem, we point this out in the discussion.

(double underline)	Word present in all of Mt, Mk and Lk
(single underline)	Word present in Mk and Mt only
(wavy underline)	Word present in Mk and Lk only
(box)	Word present in Mt and Lk only

Figure 7.1
Key to underlining

(1989). Goulder bases his commentary on the Farrer hypothesis and Gundry on the three-source hypothesis, according to both of which Luke knew and used Matthew as well as Mark. On the other hand, both Marshall and Davies and Allison assume the two-source hypothesis, according to which Matthew and Luke worked independently in their use of Mark. Assuming Markan priority, commentators on Mark need not commit themselves to a particular theory of synoptic relationships, as is the case with Hooker and with France.

It is worth noting that Gundry, in his own version of a stylometric approach, uses word-statistics to characterise Matthew's style and aspects of the way in which Matthew edits the text of Mark. Gundry's Greek Index[2] gives for each word the frequency of occurrence (i) in Matthean insertions in paralleled material, (ii) in single tradition passages peculiar to Matthew, and (iii) in occurrences shared with Mark or Luke or both. The total frequency of occurrences in Matthew, Mark and Luke, respectively, is also recorded. Goulder[3] too has a Greek vocabulary, in his case of expressions that may be regarded as characteristic of Luke, with frequencies of their occurrence in Matthew, Mark, Luke and Luke's companion volume of the *Acts of the Apostles*.

7.2 The Healing of a Leper

The Huck pericope number 45, with parallel texts Mt 8:1-4 || Mk 1:40-45 || Lk 5:12-16, that is to be examined in this section, is one that appears in Tables 3.8, 5.4 and 5.5 as a candidate for further investigation.[4] In Tables 5.4

[2]See Gundry (1994), pp. 674-682 for the Greek index and pp. 2-5 for the rationale.

[3]Goulder (1989), pp. 800-809

[4]It is worth noting that a parallel to this pericope is also to be found in the non-canonical "Unknown Gospel" fragments of *Papyrus Egerton 2* and *Papyrus Köln 255*. See Ehrman and Pleše (2011), pp. 250-251, and Watson (2013), pp. 321-324. However, we shall not make any direct use of this parallel here.

Table 7.2
Counts of verbal agreements with Mk 1:40-45

		Luke		
		0	1	total
Matthew	0	47 (31.96)	11 (26.04)	58
	1	7 (22.04)	33 (17.96)	40
	total	54	44	98

and 5.5, Mk 1:40-44 is suggested as the Markan passage to be investigated, omitting Mk 1:45 and correspondingly omitting also Lk 5:15-16.[5] We may also omit the first verse of the Matthean parallel, Mt 8:1, as this forms the transition from the Sermon on the Mount to the story of the healing of a leper that immediately follows in Matthew. This verse is not paralleled in Mark. Similarly the first part of Lk 5:12 is transitional material that is not of direct interest for our comparison with Mark. There then remain the parallel texts Mt 8:2-4 || Mk 1:40-44 || Lk 5:12b-14. However, on closer examination of the texts, it does turn out to be of interest to include Mk 1:45 and the loosely parallel Lk 5:15-16, so that in Figure 7.2 we present the parallel Greek texts Mt 8:2-4 || Mk 1:40-45 || Lk 5:12b-16 and in Figure 7.3 the corresponding parallel English texts.

In Table 7.2 we present in a contingency table the counts of verbal agreements and non-agreements aggregated over the 98 words in the section of text Mk 1:40-45, exactly as was done in Table 3.1 for the whole of Mark's gospel, with 1 the label for verbal agreements and 0 for non-agreements. As in Table 3.1, the figures in brackets below the observed frequencies are the expected frequencies, given the row and column totals, under the hypothesis that Matthew and Luke are statistically independent in their verbal agreements with Mark. For this first example we shall discuss the figures in more detail than for the later examples, but the overall conclusions will be broadly similar in all cases. We see that Matthew has retained 40 of the 98 words in Mark unchanged and Luke has retained 44 words unchanged. The expected frequencies are calculated on the basis that Matthew chose at random 40 words from the 98 words in Mark to retain unchanged, Luke chose at random

[5]The technical statistical reason for the absence of Mk 1:45 from the passages suggested in Tables 5.4 and 5.5 is that, in the decoding associated with the hidden Markov models for Matthew's use of Mark, these were State 2 passages; but there is no parallel in Matthew to Mk 1:45 and this is decoded as a State 1 passage, where Matthew is not using Mark at all.

Mt 8:2-4	Mk 1:40-45	Lk 5:12b-16
2 καὶ ἰδοὺ λεπρὸς προσελθὼν προσεκύνει αὐτῷ λέγων· κύριε, ἐὰν θέλῃς δύνασαί με καθαρίσαι. 3 καὶ ἐκτείνας τὴν χεῖρα ἥψατο αὐτοῦ λέγων· θέλω, καθαρίσθητι· καὶ εὐθέως ἐκαθαρίσθη αὐτοῦ ἡ λέπρα. 4 καὶ λέγει αὐτῷ ὁ Ἰησοῦς· ὅρα μηδενὶ εἴπῃς, ἀλλὰ ὕπαγε σεαυτὸν δεῖξον τῷ ἱερεῖ καὶ προσένεγκον τὸ δῶρον ὃ προσέταξεν Μωϋσῆς, εἰς μαρτύριον αὐτοῖς.	40 Καὶ ἔρχεται πρὸς αὐτὸν λεπρὸς παρακαλῶν αὐτὸν καὶ γονυπετῶν λέγων αὐτῷ ὅτι ἐὰν θέλῃς δύνασαί με καθαρίσαι. 41 καὶ σπλαγχνισθεὶς ἐκτείνας τὴν χεῖρα αὐτοῦ ἥψατο καὶ λέγει αὐτῷ· θέλω, καθαρίσθητι· 42 καὶ εὐθὺς ἀπῆλθεν απ᾽ αὐτοῦ ἡ λέπρα, καὶ ἐκαθαρίσθη. 43 καὶ ἐμβριμησάμενος αὐτῷ εὐθὺς ἐξέβαλεν αὐτὸν 44 καὶ λέγει αὐτῷ· ὅρα μηδενὶ μηδὲν εἴπῃς, ἀλλὰ ὕπαγε σεαυτὸν δεῖξον τῷ ἱερεῖ καὶ προσένεγκε περὶ τοῦ καθαρισμοῦ σου ἃ προσέταξεν Μωϋσῆς, εἰς μαρτύριον αὐτοῖς. 45 ὁ δὲ ἐξελθὼν ἤρξατο κηρύσσειν πολλὰ καὶ διαφημίζειν τὸν λόγον, ὥστε μηκέτι αὐτὸν δύνασθαι φανερῶς εἰς πόλιν εἰσελθεῖν, ἀλλ᾽ ἔξω ἐπ᾽ ἐρήμοις τόποις ἦν· καὶ ἤρχοντο πρὸς αὐτὸν πάντοθεν.	καὶ ἰδοὺ ἀνὴρ πλήρης λέπρας· ἰδὼν δὲ τὸν Ἰησοῦν, πεσὼν ἐπὶ πρόσωπον ἐδεήθη αὐτοῦ λέγων· κύριε, ἐὰν θέλῃς δύνασαί με καθαρίσαι. 13 καὶ ἐκτείνας τὴν χεῖρα ἥψατο αὐτοῦ λέγων· θέλω, καθαρίσθητι· καὶ εὐθέως ἡ λέπρα ἀπῆλθεν απ᾽ αὐτοῦ. 14 καὶ αὐτὸς παρήγγειλεν αὐτῷ μηδενὶ εἰπεῖν, ἀλλὰ ἀπελθὼν δεῖξον σεαυτὸν τῷ ἱερεῖ καὶ προσένεγκε περὶ τοῦ καθαρισμοῦ σου καθὼς προσέταξεν Μωϋσῆς, εἰς μαρτύριον αὐτοῖς. 15 διήρχετο δὲ μᾶλλον ὁ λόγος περὶ αὐτοῦ, καὶ συνήρχοντο ὄχλοι πολλοὶ ἀκούειν καὶ θεραπεύεσθαι ἀπὸ τῶν ἀσθενειῶν αὐτῶν· 16 αὐτὸς δὲ ἦν ὑποχωρῶν ἐν ταῖς ἐρήμοις καὶ προσευχόμενος.

Figure 7.2
The Healing of a Leper (NA[25] Greek)

Mt 8:2-4	Mk 1:40-45	Lk 5:12b-16
2 and there was a leper who came to him and knelt before him, saying, Lord, if you choose, you can make me clean. 3 He stretched out his hand and touched him, saying, I do choose. Be made clean! Immediately his leprosy was cleansed. 4 Then Jesus said to him, See that you say nothing to anyone; but go, show yourself to the priest, and offer the gift that Moses commanded, as a testimony to them.	40 A leper came to him begging him, and kneeling he said to him, If you choose, you can make me clean. 41 Moved with pity, Jesus stretched out his hand and touched him, and said to him, I do choose. Be made clean! 42 Immediately the leprosy left him, and he was made clean. 43 After sternly warning him he sent him away at once, 44 saying to him, See that you say nothing to anyone; but go, show yourself to the priest, and offer for your cleansing what Moses commanded, as a testimony to them. 45 But he went out and began to proclaim it freely, and to spread the word, so that Jesus could no longer go into a town openly, but stayed out in the country; and people came to him from every quarter.	12 there was a man covered with leprosy. When he saw Jesus, he bowed with his face to the ground and begged him, Lord, if you choose, you can make me clean. 13 Then Jesus stretched out his hand, touched him, and said, I do choose. Be made clean. Immediately the leprosy left him. 14 And he ordered him to tell no one. Go, he said, and show yourself to the priest, and, as Moses commanded, make an offering for your cleansing, for a testimony to them. 15 But now more than ever the word about Jesus spread abroad; many crowds would gather to hear him and to be cured of their diseases. 16 But he would withdraw to deserted places and pray.

Figure 7.3
The Healing of a Leper (NRSV translation)

44 words from the 98 words in Mark to retain unchanged, and the choices made by Matthew and Luke were statistically independent of each other. The expected frequencies so calculated provide a benchmark against which to compare the observed frequencies in order to examine the strength of association between the choices made by Matthew and Luke. However, as discussed in Section 2.6, we are not in a situation where we can carry out a simple significance test for association, i.e., for statistical dependence.

Just as in Table 3.1, we see again that the observed frequencies along the diagonal of the table are considerably greater than the expected frequencies, and the off-diagonal observed frequencies are less than the expected frequencies. Under the hypothesis of statistical independence, there are too many agreements between Matthew and Luke in the words that they retain unchanged (the 33 double-underlined words in the text of Mark in Figure 7.2) and in the words that they omit or alter (the 47 words without any underline in the text of Mark), and there are too few words that are retained unaltered by only one of Matthew or Luke (the $7 + 11 = 18$ words with a single underline, either straight or wavy). On the face of it, there appears to be strong evidence that Matthew and Luke were not independent in their choice of words to retain unchanged from Mark, but rather that there was a positive association between their choices.

Although the main body of our earlier statistical analysis, in Chapters 3 and 5, and the resulting conclusions did not depend in any way upon a direct comparison of the texts of Matthew and Luke, it is in addition worth noting from Figure 7.2 the 4 boxed words which are the minor verbal agreements of Matthew and Luke against Mark in the specific sense of words that are present in both Matthew and Luke, but absent from Mark. These minor agreements may be argued to provide further evidence beyond that of our main statistical analysis, and beyond that of the data summarised in our contingency table, Table 7.2, that Matthew and Luke did not work independently of each other.

In our discussion of the minor agreements in Section 6.6, we focussed on the positive agreements, specifically the minor verbal agreements of Matthew and Luke against Mark, which were common insertions and common changes with respect to the text of Mark. Now, and in particular in our consideration of the contingency tables associated with our texts, we are more concerned with other types of agreement. In his classification of the minor agreements into different types, Vinson (2004) included the category of "common omissions", which Neirynck (1974)[6] had referred to as "agreements in omission" or "negative agreements". In our counts they are included among the words in Mark that are absent from both Matthew and Luke and hence contribute to the observed frequency of 47 in the top left-hand corner of Table 7.2, and similarly elsewhere in our other contingency tables. However, our considerations here go beyond such a classification of minor agreements. In particular, we also take into consideration what we may term "common retentions", where the texts of

[6]Neirynck (1974), p. 52

Matthew and Luke agree in what words of Mark are retained unchanged. The count of such common retentions appears as the observed frequency of 33 in the bottom right-hand corner of Table 7.2. Commentators, in their discussions of the use of Mark by Matthew and Luke, have considered in detail the various forms of minor agreement, but relatively little attention has been paid to the extent of the common retentions, which is an important aspect of the approach presented here.

After these statistical observations, we turn to a more detailed examination of the texts themselves to see what are the specific features that are reflected in the statistics. We shall look at a selection of the most noteworthy issues without trying to be comprehensive in examining all verbal agreements and non-agreements. To begin with the common retentions, there are some notable phrases, central to the plot, that are word-for-word common to all three synoptic gospels: the leper says "if you choose, you can make me clean"; in response Jesus "stretched out his hand ... touched", saying "I do choose. Be made clean!"; after the healing Jesus instructs the man "go, show yourself to the priest ... Moses commanded ... a testimony to them". The reasons for the occurrence of precisely these common retentions are not discussed in the commentaries, and it might be the case that Matthew and Luke independently decided to retain these sections of the Markan text word for word. However, it is easier to explain these common retentions if we suppose, for example, that Luke knew both Mark and Matthew, in which case the presence of exactly the same text in both his sources would make it more likely that Luke also would retain the text unchanged.

Striking too are some of the common omissions. Although most of Mk 1:41 is word-for-word present in Matthew and Luke, the word σπλαγχνισθεὶς ("moved with pity") is present in neither, a fact that is highlighted in the commentaries. One possible explanation might be that both Matthew and Luke independently wished to avoid attributing emotion to Jesus, but, as pointed out by Goulder,[7] Luke uses the same verb at Lk 7:13 in his single-tradition account of the raising of the widow's son at Nain. Matthew too uses the same verb at Mt 9:36, 14:14, 15:32, 20:34 of Jesus having compassion on the crowds and in the healing of two blind men, in the last case specifically inserting it,[8] so it appears that Matthew and Luke are not totally averse to expressing the compassion of Jesus. As Davies and Allison[9] put it, "One would certainly be hard pressed to urge that both evangelists [independently] omitted σπλαγχνισθεὶς". But Davies and Allison point to a different explanation, as do both Hooker and France.[10] They suggest that the original reading in Mark was ὀργισθεὶς ("moved with anger" — most plausibly, anger at the evil of disease), as found in a minority of manuscripts, instead of σπλαγχνισθεὶς. This more difficult reading, showing Jesus as angry, would be more likely to have been

[7] Goulder (1989), p. 329

[8] See Section 7.10.

[9] Davies and Allison (1997), Volume II, p. 13

[10] Hooker (1991), p. 79; France (2002), p. 115

found inappropriate by both Matthew and Luke and therefore omitted. So here we have an example of textual corruption as a proposed explanation for a minor agreement.

In the same vein, the whole of the verse Mk 1:43 is absent from both Matthew and Luke. This verse too may be regarded as strongly emotive and excessively harsh, since the word ἐμβριμησάμενος ("sternly warning") can be used as an expression of anger, and may be so understood here. Furthermore, ἐξέβαλεν ("sent away") is more commonly used in the stronger sense of driving out, or for the casting out of demons. Both Matthew and Luke may have wished to avoid the attribution of such strong emotions of anger to Jesus. So on these grounds it is possible to argue that these omissions were made independently by Matthew and Luke. On the other hand, a simpler explanation is available if it is assumed, as do Goulder and Gundry,[11] that Luke had both Mark and Matthew as sources and here followed Matthew in his omissions.

In Mk 1:44 there is an emphatic negative μηδενὶ μηδὲν ("to no one nothing"), but both Matthew and Luke omit μηδὲν. The assumption that Luke had Matthew as a source, or vice versa, makes this easier to explain than if they worked independently.[12]

A further omission by Matthew is that of the whole of the verse Mk 1:45, in which the leper, against Jesus' instructions, spreads the news of his healing, with the result that Jesus can no longer enter any town because of the crowds. Gundry[13] suggests that Matthew wishes to avoid the apparent contradiction with the next verse in both Matthew and Mark, in which Jesus returns to the town of Capernaum. Davies and Allison,[14] however, suggest that the omission occurs because Matthew does not want to portray the authority of Jesus as being undermined. This suggestion also accounts for Luke's handling of this verse. In Lk 5:15-16 he replaces Mk 1:45 with material in which news about Jesus is spread abroad, but not through the agency of the leper. As Marshall[15] puts it: "The command to silence is retained, but Luke is unwilling to attribute direct disobedience of it to the cured leper." There are in Mk 1:45 || Lk 5:15-16, according to our criteria, four verbal agreements between Mark and Luke, but overall the parallel is very loose. It is perhaps debatable whether, with regard to this verse in Mark, it is more straightforward to envisage Matthew and Luke working independently of each other or whether it is easier to envisage Luke having both Mark and Matthew as sources and, perhaps encouraged by the absence of this verse in Matthew, feeling free to replace Mark's text with something substantially different.

[11] Goulder (1989), p. 329; Gundry (1994), p. 139

[12] As an issue in textual criticism, it may be noted that μηδὲν is also missing from a number of Markan manuscripts. It might be that this represents an earlier reading and that μηδὲν was not present in whatever texts of Mark were used by Matthew and Luke. However, a perhaps more likely explanation of the absence of μηδὲν in some manuscripts of Mark is that it is due to the later assimilation of the text of Mark to those of Matthew and Luke.

[13] Gundry (1994), p. 140

[14] Davies and Allison (1997), Volume II, p. 16

[15] Marshall (1978), p. 207

There are also the positive minor agreements of Matthew and Luke against Mark. Into Mk 1:40 both Matthew and Luke insert ἰδοὺ (translated "behold" in the King James Version but left untranslated in the NRSV) and κύριε ("Lord", in the vocative form). According to Gundry,[16] the insertion of ἰδοὺ and κύριε are both characteristic of Matthew's style: referring to Gundry's statistics, there are 34 Matthean insertions of ἰδοὺ in paralleled material with 9 occurrences in passages peculiar to Matthew, and 34 insertions of κύριος in paralleled material with 15 occurrences in passages peculiar to Matthew. For Gundry, Luke's insertion of these words suggests that Luke knew Matthew as well as Mark. For Davies and Allison,[17] it appears not to be surprising that Matthew and Luke independently insert ἰδοὺ and κύριε, words which occur much more frequently in Matthew and Luke than in Mark. Marshall[18] does concede that Luke rarely adds titles to his Markan source, but resorts to the suggestion that the insertion of κύριε may be due to the influence of oral sources.

The καὶ λέγει αὐτῷ ("and says to him" — using a historic present indicative) in Mk 1:41 is replaced in both Matthew and Luke by λέγων ("saying" — a present participle). This may be regarded as a stylistic improvement, which could have been carried out independently by both Matthew and Luke, but which is more easily explained if it is assumed that Luke knew Matthew, or vice versa. The εὐθὺς ("immediately") in Mk 1:42 is replaced in both Matthew and Luke by the synonymous εὐθέως. The word εὐθὺς is common in Mark, but infrequently used in Matthew and Luke, who prefer εὐθέως. So here it may reasonably be argued that Matthew and Luke both independently replace εὐθὺς by their preferred εὐθέως.

Finally, it is worth noting what in the terminology of Neirynck (1974) is an "agreement in inverted order". This is a form of minor agreement of Matthew and Luke against Mark in which the words of Mark are retained unchanged, except that their order is rearranged in exactly the same way by both Matthew and Luke. In the simplest cases, as in the following example, two adjacent words are interchanged. There is full verbal agreement, as we have defined it, in these words among all three synoptic gospels, so that this form of minor agreement is not picked up in our statistics of verbal agreements, and the words are simply double-underlined in the parallel texts. Here, in Mk 1:41 we have ἐκτείνας τὴν χεῖρα αὐτοῦ ἥψατο (literally "having stretched out his hand touched [him]", with "him" understood), whereas in Mt 8:3 and Lk 5:13 we have ἐκτείνας τὴν χεῖρα ἥψατο αὐτοῦ (literally "having stretched out [his] hand touched him", with "his" understood), where αὐτοῦ ("his"/"him") has been interchanged with ἥψατο ("touched"). This inversion of order places more emphasis on the fact that it is the ritually unclean leper who has been touched. For Davies and Allison,[19] the verb ἥψατο cries out for an object, and

[16]Gundry (1994), p. 139
[17]Davies and Allison (1997), Volume II, pp. 10, 12
[18]Marshall (1978), p. 209
[19]Davies and Allison (1997), Volume II, p. 13

so "Matthew and Luke have, naturally enough, independently moved αὐτοῦ so that it no longer qualifies 'hand' but instead becomes the object of 'touch' ". On the other hand, for Gundry,[20] Luke follows Matthew.

We see then that, overall, there are a number of common retentions of Markan material and also a variety of agreements of various kinds of Matthew and Luke against Mark, including omissions, changes and additions, which the statistics of Table 7.2 indicate are substantially more than would occur by chance. Arguments to explain the common retentions are lacking in the commentaries. In each case of an agreement against Mark, an explanation may be found for why Matthew and Luke might have independently of each other altered the text of Mark in the same way. These explanations are of different kinds, and some are more convincing than others, but in each case a simpler explanation is available if it is assumed that either Luke had Matthew as a source or vice versa. From this first section of text that we have examined in detail, we can already begin to see that advocates of the hypothesis that Matthew and Luke were independent in their use of Mark require a variety of strategies and considerable ingenuity to defend their position.

7.3 The Healing of the Paralytic

The Huck pericope number 52, with parallel texts Mt 9:1-8 || Mk 2:1-12 || Lk 5:17-26, is one that appears in Tables 5.6, 5.7 and 6.12 as a candidate for further investigation. This is the pericope that in Mark and Luke immediately follows on from the pericope of the Healing of a Leper that was studied in Section 7.2. In Tables 5.6 and 5.7 it is more specifically the passage Mk 2:8-12, approximately the second half of pericope number 52, that is highlighted. The corresponding parallel Greek texts Mt 9:4-8 || Mk 2:8-12 || Lk 5:22-26 are laid out in Figure 7.4 and the parallel English translations in Figure 7.5. It is these texts that we shall scrutinise in this section. In the earlier part of the pericope, the paralytic has been brought to Jesus, who has declared to him that his sins are forgiven, which the scribes regard as blasphemous, because it is God alone who can forgive sins. It is with Jesus reading their thoughts that our text begins.

In Table 7.3 we present the counts of verbal agreements and non-agreements aggregated over the 95 words in the section of text Mk 2:8-12 and in brackets the expected frequencies, given the row and column totals, under the hypothesis that Matthew and Luke are statistically independent in their verbal agreements with Mark. Again we see that the observed frequencies along the diagonal of the table are much greater than the expected frequencies, which again at least suggests that there is strong evidence that

[20]Gundry (1994), p. 139

Mt 9:4-8	Mk 2:8-12	Lk 5:22-26
4 καὶ εἰδὼς ὁ Ἰησοῦς τὰς ἐνθυμήσεις αὐτῶν εἶπεν· ἱνατί ἐνθυμεῖσθε πονηρὰ ἐν ταῖς καρδίαις ὑμῶν; 5 τί γάρ ἐστιν εὐκοπώτερον, εἰπεῖν· ἀφίενταί σου αἱ ἁμαρτίαι, ἢ εἰπεῖν· ἔγειρε καὶ περιπάτει; 6 ἵνα δὲ εἰδῆτε ὅτι ἐξουσίαν ἔχει ὁ υἱὸς τοῦ ἀνθρώπου ἐπὶ τῆς γῆς ἀφιέναι ἁμαρτίας – τότε λέγει τῷ παραλυτικῷ· ἔγειρε ἆρον σου τὴν κλίνην καὶ ὕπαγε εἰς τὸν οἶκόν σου. 7 καὶ ἐγερθεὶς ἀπῆλθεν εἰς τὸν οἶκον αὐτοῦ. 8 ἰδόντες δὲ οἱ ὄχλοι ἐφοβήθησαν καὶ ἐδόξασαν τὸν θεὸν τὸν δόντα ἐξουσίαν τοιαύτην τοῖς ἀνθρώποις.	8 καὶ εὐθὺς ἐπιγνοὺς ὁ Ἰησοῦς τῷ πνεύματι αὐτοῦ ὅτι οὕτως διαλογίζονται ἐν ἑαυτοῖς λέγει αὐτοῖς· τί ταῦτα διαλογίζεσθε ἐν ταῖς καρδίαις ὑμῶν; 9 τί ἐστιν εὐκοπώτερον, εἰπεῖν τῷ παραλυτικῷ· ἀφίενταί σου αἱ ἁμαρτίαι, ἢ εἰπεῖν· ἔγειρε καὶ ἆρον τὸν κράβαττόν σου καὶ περιπάτει; 10 ἵνα δὲ εἰδῆτε ὅτι ἐξουσίαν ἔχει ὁ υἱὸς τοῦ ἀνθρώπου ἀφιέναι ἁμαρτίας ἐπὶ τῆς γῆς – λέγει τῷ παραλυτικῷ· 11 σοὶ λέγω, ἔγειρε ἆρον τὸν κράβαττόν σου καὶ ὕπαγε εἰς τὸν οἶκόν σου. 12 καὶ ἠγέρθη καὶ εὐθὺς ἄρας τὸν κράβαττον ἐξῆλθεν ἔμπροσθεν πάντων, ὥστε ἐξίστασθαι πάντας καὶ δοξάζειν τὸν θεὸν λέγοντας ὅτι οὕτως οὐδέποτε εἴδαμεν.	22 ἐπιγνοὺς δὲ ὁ Ἰησοῦς τοὺς διαλογισμοὺς αὐτῶν ἀποκριθεὶς εἶπεν πρὸς αὐτούς· τί διαλογίζεσθε ἐν ταῖς καρδίαις ὑμῶν; 23 τί ἐστιν εὐκοπώτερον, εἰπεῖν· ἀφέωνταί σοι αἱ ἁμαρτίαι σου, ἢ εἰπεῖν· ἔγειρε καὶ περιπάτει; 24 ἵνα δὲ εἰδῆτε ὅτι ὁ υἱὸς τοῦ ἀνθρώπου ἐξουσίαν ἔχει ἐπὶ τῆς γῆς ἀφιέναι ἁμαρτίας – εἶπεν τῷ παραλελυμένῳ· σοὶ λέγω, ἔγειρε καὶ ἄρας τὸ κλινίδιόν σου πορεύου εἰς τὸν οἶκόν σου. 25 καὶ παραχρῆμα ἀναστὰς ἐνώπιον αὐτῶν, ἄρας ἐφ᾽ ὃ κατέκειτο, ἀπῆλθεν εἰς τὸν οἶκον αὐτοῦ δοξάζων τὸν θεόν. 26 καὶ ἔκστασις ἔλαβεν ἅπαντας καὶ ἐδόξαζον τὸν θεὸν καὶ ἐπλήσθησαν φόβου λέγοντες ὅτι εἴδομεν παράδοξα σήμερον.

Figure 7.4
The Healing of the Paralytic (NA^{25} Greek)

Mt 9:4-8	Mk 2:8-12	Lk 5:22-26
4 But Jesus, perceiving their thoughts, said, "Why do you think evil in your hearts? 5 For which is easier, to say, 'Your sins are forgiven,' or to say, 'Stand up and walk'? 6 But so that you may know that the Son of Man has authority on earth to forgive sins"—he then said to the paralytic—"Stand up, take your bed and go to your home." 7 And he stood up and went to his home. 8 When the crowds saw it, they were filled with awe, and they glorified God, who had given such authority to human beings.	8 At once Jesus perceived in his spirit that they were discussing these questions among themselves; and he said to them, "Why do you raise such questions in your hearts? 9 Which is easier, to say to the paralytic, 'Your sins are forgiven,' or to say, 'Stand up and take your mat and walk'? 10 But so that you may know that the Son of Man has authority on earth to forgive sins—he said to the paralytic— 11 "I say to you, stand up, take your mat and go to your home." 12 And he stood up, and immediately took the mat and went out before all of them; so that they were all amazed and glorified God, saying, "We have never seen anything like this!"	22 When Jesus perceived their questionings, he answered them, "Why do you raise such questions in your hearts? 23 Which is easier, to say, 'Your sins are forgiven you,' or to say, 'Stand up and walk'? 24 But so that you may know that the Son of Man has authority on earth to forgive sins"—he said to the one who was paralyzed—"I say to you, stand up and take your bed and go to your home." 25 Immediately he stood up before them, took what he had been lying on, and went to his home, glorifying God. 26 Amazement seized all of them, and they glorified God and were filled with awe, saying, "We have seen strange things today."

Figure 7.5
The Healing of the Paralytic (NRSV translation)

Table 7.3
Counts of verbal agreements with Mk 2:8-12

		Luke		
		0	1	total
Matthew	0	37 (19.92)	7 (24.08)	44
	1	6 (23.08)	45 (27.92)	51
	total	43	52	95

Matthew and Luke were not statistically independent in their choice of words to retain unchanged from Mark, but that there was a positive association between their choices. It is in addition worth noting from Figure 7.4 the 7 boxed words which are the minor verbal agreements of Matthew and Luke against Mark.

On a more detailed examination of the texts, we may observe first of all the long sequences of common retentions by Matthew and Luke in the block of material from the last few words of Mk 2:8, ἐν ταῖς καρδίαις ὑμῶν ("in your hearts"), up to and including Mk 2:11, but broken up mainly by some notable common omissions. The words τῷ παραλυτικῷ ("to the paralytic") from Mk 2:9 and καὶ ἆρον τὸν κράβαττόν σου ("and take your mat") are omitted by both Matthew and Luke. This has the effect of making Jesus' comment of more general applicability than to the case of the paralytic that is immediately at hand. For Davies and Allison[21] it is not surprising that Matthew and Luke should have made these omissions independently of each other. However, the common omission is more readily accounted for if we suppose that Matthew decided to omit these words and then Luke followed suit, or vice versa.

The omission or replacement by Matthew and Luke of Mark's τὸν κράβαττόν ("the mat") at Mk 2:9,11,12, and also in the first half of the pericope at Mk 2:4, has led to considerable discussion. The word κράβαττος[22] is a colloquial term for a poor man's bed — a mat or mattress or pallet. For Streeter,[23] it is more or less inevitable that Matthew and Luke, even working independently, would correct this "vulgarism". Matthew at Mt 9:2,6 has replaced it by the more literary word κλίνη, whose primary meaning is bed or couch, but which may also be used for a pallet or stretcher. Luke replaces κράβαττος by κλίνη at Lk 5:18, by its diminutive κλινίδιον at 5:19,24 and by

[21] Davies and Allison (1997), Volume II, p. 92

[22] This is the nominative form. The alternative spellings κράββατος and κράβατος are also found, the latter in NA[25], but we have kept to κράβαττος, as used in NA[28].

[23] Streeter (1924), p. 299

the phrase ἐφ' ὃ κατέκειτο ("what he had been lying on") at 5:25, a phrase similar to one used by Mark at Mk 2:4.[24] The word κράβαττος appears in neither Matthew's nor Luke's gospel, although Luke does use it twice in the *Acts of the Apostles*, at Acts 5:15, 9:33, so that we may conclude that Luke is not altogether averse to its use. Because of this, Goulder[25] argues that, if Luke knew only Mark, there would have been no reason for him to alter the word κράβαττος. But if he also knew Matthew then he was influenced by Matthew's replacement of κράβαττος by κλίνη. However, where Luke uses κλίνη elsewhere in his gospel, it is in the sense of a proper bed, and he is rather hesitant about using this word for a simple form of bedding. Hence his use here not only of κλίνη but also of its diminutive and the phrase ἐφ' ὃ κατέκειτο. The argument based on such considerations for Luke's knowledge of Matthew is given in Goulder (1978) but countered as inconclusive in Tuckett (1984).

There is also an agreement in inverted order of Matthew and Luke against Mark. Whereas at Mk 2:10 we find ἐξουσίαν ἔχει ὁ υἱὸς τοῦ ἀνθρώπου ἀφιέναι ἁμαρτίας ἐπὶ τῆς γῆς ("the Son of Man has authority on earth to forgive sins"), at both Mt 9:6 and Lk 5:24 the words ἐπὶ τῆς γῆς ("on earth") have been moved to a position immediately after ὁ υἱὸς τοῦ ἀνθρώπου ("the Son of Man"). The English translation given in Figure 7.5 is the same in either case, but, as pointed out by Gundry,[26] the move does imply a change of stress from the earth as the place of forgiveness to the earth as the place where Jesus, the Son of Man, exercises his authority. Again the question arises as to how plausible it is that Matthew and Luke should have made this change independently of each other.

Furthermore there is the important common insertion by Matthew at Mt 9:7 and Luke at Lk 5:25 into the text of Mk 8:12 of the words ἀπῆλθεν εἰς τὸν οἶκον αὐτοῦ ("went to his home"). In the previous verse, Mt 9:6 || Mk 8:11 || Lk 5:24, Jesus has commanded the paralytic to stand up, take his bed/mat and go home. Mark then reports that the man stood up, took his mat and went out, but Matthew and Luke, using precisely the same words, emphatically add that the man went home, exactly following Jesus' instructions, which Mark presumably takes for granted. This insertion is particularly noteworthy, because generally Matthew and Luke have a tendency to abbreviate Mark's narratives rather than to expand them. Yet again, the question that arises is how plausible it is that Matthew and Luke would have made the insertion independently of each other, or is it really much more likely that one of them did so and then the other, following, agreed with him.

In the last verse of this pericope, Mt 9:8 || Mk 8:12 || Lk 5:26, both Matthew and Luke add to Mark the comment that the onlookers "were filled with

[24]As a comment on how verbal agreements are counted, we may note that the word κατέκειτο appears at both Mk 2:4 and Lk 5:25, in the same pericope. However, this has not been counted by Farmer as a verbal agreement, as the two occurrences have apparently been judged as being too far apart in context.

[25]Goulder (1989), p. 332

[26]Gundry (1994), p. 164

awe", although the same words in the English translation mask the fact that different, though related, words are used in the Greek, with the noun φόβος ("fear", "awe") in Luke and the cognate verb φοβέω in Matthew. Again we have a positive agreement, though a looser one with no exact verbal agreement, of Matthew and Luke against Mark.

Overall, just as in the example of Section 7.2, it is easier to explain the various agreements of Matthew and Luke against Mark by assuming that Matthew was also a source for Luke, or vice versa, instead of assuming that they were working independently of each other.

7.4 The Feeding of the Five Thousand

Our next example is the Huck pericope number 112 with parallel texts Mt 14:13-21 || Mk 6:30-44 || Lk 9:10-17, which takes us to a later point in the gospels and the feeding of the five thousand. This pericope appears in Tables 5.4, 5.5 and 6.12 as a candidate for further investigation. In Tables 5.4 and 5.5 it is specifically the passage Mk 6:37-44 that is selected, which makes up the second half of the Huck pericope number 112 and which we shall examine here. The parallel Greek texts Mt 14:16-21 || Mk 6:37-44 || Lk 9:13-17 are laid out in Figure 7.6 and the parallel English texts in Figure 7.7. (The last word βρώματα of Mt 14:15 is included in Figure 7.6 to account for the minor verbal agreement shown by the box around the same word in Lk 9:13. Similarly, the word "food" from Mt 14:15 is included in the English version in Figure 7.7.) Just before the point where the story is taken up in our selected passage, the twelve disciples have suggested to Jesus that he should send the people away so that they can buy food in the surrounding villages, but, at the beginning of our text, Jesus challenges his disciples to feed the people themselves.

In Table 7.4 we present the counts of verbal agreements aggregated over the 111 words in the section of text Mk 6:37-44 and in brackets the expected frequencies, given the row and column totals, under the hypothesis that Matthew and Luke are statistically independent in their verbal agreements with Mark. Yet again we see that the observed frequencies along the diagonal of the table are much greater than the expected frequencies. This suggests that there is very strong evidence that Matthew and Luke were not statistically independent in their choice of words to retain unchanged from Mark, but with the usual caveat that we may not validly carry out a simple chi-square test of significance. It is also worth noting from Figure 7.6 the several instances, 9 boxed words in total, of minor verbal agreements of Matthew and Luke against Mark.

The feeding of the five thousand is unique among the miracles of Jesus in that it appears in all four of the canonical gospels, including John, although it is only the synoptic gospels that we are concerned with here. Furthermore,

Mt 14:16-21	Mk 6:37-44	Lk 9:13-17
βρώματα. 16 ὁ δὲ Ἰησοῦς εἶπεν αὐτοῖς· οὐ χρείαν ἔχουσιν ἀπελθεῖν, δότε αὐτοῖς ὑμεῖς φαγεῖν. 17 οἱ δὲ λέγουσιν αὐτῷ· οὐκ ἔχομεν ὧδε εἰ μὴ πέντε ἄρτους καὶ δύο ἰχθύας. 18 ὁ δὲ εἶπεν· φέρετέ μοι ὧδε αὐτούς. 19 καὶ κελεύσας τοὺς ὄχλους ἀνακλιθῆναι ἐπὶ τοῦ χόρτου, λαβὼν τοὺς πέντε ἄρτους καὶ τοὺς δύο ἰχθύας, ἀναβλέψας εἰς τὸν οὐρανὸν εὐλόγησεν καὶ κλάσας ἔδωκεν τοῖς μαθηταῖς τοὺς ἄρτους, οἱ δὲ μαθηταὶ τοῖς ὄχλοις. 20 καὶ ἔφαγον πάντες καὶ ἐχορτάσθησαν, καὶ ἦραν τὸ περισσεῦον τῶν κλασμάτων δώδεκα κοφίνους πλήρεις. 21 οἱ δὲ ἐσθίοντες ἦσαν ἄνδρες ὡσεὶ πεντακισχίλιοι χωρὶς γυναικῶν καὶ παιδίων.	37 ὁ δὲ ἀποκριθεὶς εἶπεν αὐτοῖς· δότε αὐτοῖς ὑμεῖς φαγεῖν. καὶ λέγουσιν αὐτῷ· ἀπελθόντες ἀγοράσωμεν δηναρίων διακοσίων ἄρτους καὶ δώσομεν αὐτοῖς φαγεῖν; 38 ὁ δὲ λέγει αὐτοῖς· πόσους ἔχετε ἄρτους; ὑπάγετε ἴδετε. καὶ γνόντες λέγουσιν· πέντε, καὶ δύο ἰχθύας. 39 καὶ ἐπέταξεν αὐτοῖς ἀνακλιθῆναι πάντας συμπόσια συμπόσια ἐπὶ τῷ χλωρῷ χόρτῳ. 40 καὶ ἀνέπεσαν πρασιαὶ πρασιαὶ κατὰ ἑκατὸν καὶ κατὰ πεντήκοντα. 41 καὶ λαβὼν τοὺς πέντε ἄρτους καὶ τοὺς δύο ἰχθύας ἀναβλέψας εἰς τὸν οὐρανὸν εὐλόγησεν καὶ κατέκλασεν τοὺς ἄρτους καὶ ἐδίδου τοῖς μαθηταῖς ἵνα παρατιθῶσιν αὐτοῖς, καὶ τοὺς δύο ἰχθύας ἐμέρισεν πᾶσιν. 42 καὶ ἔφαγον πάντες καὶ ἐχορτάσθησαν, 43 καὶ ἦραν κλάσματα δώδεκα κοφίνων πληρώματα καὶ ἀπὸ τῶν ἰχθύων. 44 καὶ ἦσαν οἱ φαγόντες τοὺς ἄρτους πεντακισχίλιοι ἄνδρες.	13 εἶπεν δὲ πρὸς αὐτούς· δότε αὐτοῖς φαγεῖν ὑμεῖς. οἱ δὲ εἶπαν· οὐκ εἰσὶν ἡμῖν πλεῖον ἢ ἄρτοι πέντε καὶ ἰχθύες δύο, εἰ μήτι πορευθέντες ἡμεῖς ἀγοράσωμεν εἰς πάντα τὸν λαὸν τοῦτον βρώματα. 14 ἦσαν γὰρ ὡσεὶ ἄνδρες πεντακισχίλιοι. εἶπεν δὲ πρὸς τοὺς μαθητὰς αὐτοῦ· κατακλίνατε αὐτοὺς κλισίας ὡσεὶ ἀνὰ πεντήκοντα. 15 καὶ ἐποίησαν οὕτως καὶ κατέκλιναν ἅπαντας. 16 λαβὼν δὲ τοὺς πέντε ἄρτους καὶ τοὺς δύο ἰχθύας ἀναβλέψας εἰς τὸν οὐρανὸν εὐλόγησεν αὐτοὺς καὶ κατέκλασεν καὶ ἐδίδου τοῖς μαθηταῖς παραθεῖναι τῷ ὄχλῳ. 17 καὶ ἔφαγον καὶ ἐχορτάσθησαν πάντες, καὶ ἤρθη τὸ περισσεῦσαν αὐτοῖς κλασμάτων κόφινοι δώδεκα.

Figure 7.6
The Feeding of the Five Thousand (NA^{25} Greek)

Mt 14:16-21	Mk 6:37-44	Lk 9:13-17
. . . food . . . 16 Jesus said to them, "They need not go away; you give them something to eat." 17 They replied, "We have nothing here but five loaves and two fish." 18 And he said, "Bring them here to me." 19 Then he ordered the crowds to sit down on the grass. Taking the five loaves and the two fish, he looked up to heaven, and blessed and broke the loaves, and gave them to the disciples, and the disciples gave them to the crowds. 20 And all ate and were filled; and they took up what was left over of the broken pieces, twelve baskets full. 21 And those who ate were about five thousand men, besides women and children.	37 But he answered them, "You give them something to eat." They said to him, "Are we to go and buy two hundred denarii worth of bread, and give it to them to eat?" 38 And he said to them, "How many loaves have you? Go and see." When they had found out, they said, "Five, and two fish." 39 Then he ordered them to get all the people to sit down in groups on the green grass. 40 So they sat down in groups of hundreds and of fifties. 41 Taking the five loaves and the two fish, he looked up to heaven, and blessed and broke the loaves, and gave them to his disciples to set before the people; and he divided the two fish among them all. 42 And all ate and were filled; 43 and they took up twelve baskets full of broken pieces and of the fish. 44 Those who had eaten the loaves numbered five thousand men.	13 But he said to them, "You give them something to eat." They said, "We have no more than five loaves and two fish—unless we are to go and buy food for all these people." 14 For there were about five thousand men. And he said to his disciples, "Make them sit down in groups of about fifty each." 15 They did so and made them all sit down. 16 And taking the five loaves and the two fish, he looked up to heaven, and blessed and broke them, and gave them to the disciples to set before the crowd. 17 And all ate and were filled. What was left over was gathered up, twelve baskets of broken pieces.

Figure 7.7
The Feeding of the Five Thousand (NRSV translation)

Table 7.4
Counts of verbal agreements with Mk 6:37-44

		Luke		
		0	1	total
Matthew	0	60 (41.58)	5 (23.42)	65
	1	11 (29.42)	35 (16.58)	46
	total	71	40	111

as Hooker[27] reminds us, it is possible to take the miracle of the feeding of the four thousand at Mt 15:32-39 || Mk 8:1-10 as a doublet, a reminiscence of the same event, which would give us two more accounts. Whether or not we accept such a view, this was clearly a miracle of considerable importance for early Christian communities. It has many resonances with other biblical miracles of feeding and accounts of communal meals, but above all it foreshadows the Last Supper and the institution of the Christian Eucharist.

The pericope as a whole has been recognised as a crucial one for the study of the minor agreements. Thus Streeter writes:[28] "The Feeding of the Five Thousand is a section in which there are more minor agreements than in any other of the same length", though this raises the question of exactly how the minor agreements are specified and counted by Streeter. Marshall[29] somewhat cursorily, without presenting any detailed argument, dismisses any agreements between Matthew and Luke against Mark as mainly mere coincidence and otherwise by accounting for them by an appeal to oral tradition: "Luke's narrative is closely dependent upon that of Mk.; there are one or two agreements with Mt. against Mk. which may go beyond mere coincidence, but they hardly testify to anything more than the existence of continuing oral traditions of a familiar story." If oral tradition is to be invoked to account for the agreements of Matthew and Luke against Mark, it must presumably be supposed that Matthew and Luke were influenced by similar oral traditions that were not available to Mark or not used by him, but it is not at all clear how likely such a scenario might be. Davies and Allison[30] examine the minor agreements in more detail and argue that "all the minor agreements can be readily explained in terms of redactional tendencies" to conclude that "the minor agreements do not require modification of the standard two-source theory." So Davies and

[27]Hooker (1991), p. 163
[28]Streeter (1924), p. 313
[29]Marshall (1978), p. 358
[30]Davies and Allison (1997), Volume II, pp. 478-480

Allison appeal to the use of similar editorial policies by Matthew and Luke rather than exposure to similar oral tradition. Gundry,[31] on the other hand, as we should expect, argues that the minor agreements imply Matthean influence on Luke. He writes of Luke: "...attention shows a series of coincidences with the Matthean version that cannot be accidental. These are in fact so notorious that we must consider the possibility of Luke's turning up the story in Matthew as well as in Mark before writing." We have here then a striking example of where New Testament scholars, from the same evidence, arrive at altogether opposite conclusions.

We turn now to a more detailed examination of the text, beginning with the common retentions, on which the commentators place relatively little emphasis. The numerical details have been retained by both Matthew and Luke, and also by John, that there were five loaves and two fish, five thousand men were fed, and there were twelve baskets of leftovers. According to France,[32] "The point of the numbers is ...to emphasise the magnitude of the miracle ...as such they would naturally become fixed in the oral repetition of the story." It should then perhaps not be surprising that we find them in all four gospels. We focus on three more sustained sections of word-for-word agreement in the Greek text. The first is what may be regarded as the focal point of the narrative at Mt 14:19 || Mk 6:41 || Lk 9:16, that

> λαβὼν τοὺς πέντε ἄρτους καὶ τοὺς δύο ἰχθύας ἀναβλέψας εἰς τὸν οὐρανὸν εὐλόγησεν ... τοῖς μαθηταῖς
>
> Taking the five loaves and the two fish, he looked up to heaven, and blessed ...to the disciples.

What Jesus is doing here is, from one point of view, what any Jewish head of household or host would do in giving thanks before a meal. However, the words of Jesus at the last supper, at Mt 26:26 || Mk 14:22 || Lk 22:19, are being echoed — and the words at the celebration of the Christian Eucharist, which in the setting of some form of early Christian liturgy would presumably have been known to the gospel writers. At the Last Supper Jesus gives the bread and wine to the disciples, and here too it is to the disciples that he gives the bread and fish, to be distributed to the people. In this context, the use of the four verbs, take, bless, break, give, is hardly surprising, and so the word-for-word common retention seems not implausible, even if Matthew and Luke were using Mark independently.

---

[31] Gundry (1994), p. 433

[32] France (2002), p. 266. For Dunn (2013), p. 150, too, the numbers are the fixed points in the oral tradition around which the performer would elaborate.

Another common retention occurs at the beginning of our text at Mt 14:16 || Mk 6:37 || Lk 9:13, where Jesus says to his disciples

> δότε αὐτοῖς ὑμεῖς φαγεῖν
>
> "You give them something to eat."[33]

When in the surrounding text both Matthew and Luke have made many alterations and omissions, it is not at all clear why, if they were working independently, they should both have decided to retain unchanged precisely these words. Thirdly, there is the common retention towards the end of the pericope at Mt 14:20 || Mk 6:42 || Lk 9:17,

> καὶ ἔφαγον πάντες καὶ ἐχορτάσθησαν
>
> And all ate and were filled

with the minor difference in Luke that he moves πάντες ("all") to the end of the clause to emphasise that *all* were filled. Again, if Matthew and Luke were working independently, it is not clear why they should both have decided to retain exactly these words.

Turning to the common omissions, an important example occurs where both Matthew and Luke omit much of the dialogue between Jesus and his disciples in Mk 6:37b-38a. The commentators note the apparent irony in the disciples' reply at Mk 6:37b, which may be considered subversive of Jesus' authority — or perhaps the disciples are simply being obtuse. Mark is generally not afraid of portraying the disciples' total lack of understanding, but Matthew and Luke do not wish to draw so negative a portrait of the disciples. For these reasons it appears plausible that Matthew and Luke might have decided, independently of each other, to omit this material.[34] But then Matthew and Luke both also omit much of the detail of the seating arrangements in Mk 6:39-40. In discussing Mark's text, France[35] writes: "The vivid description suggests the eyewitness account of someone who was present at this extraordinary picnic." Now Matthew and Luke both decide to truncate Mark's account here, much more apparent in the Greek text than in the English translation, so losing this vividness of detail. Furthermore, having then followed Mark for

---

[33] According to the NA^{25} Greek of Luke, there is an inversion of order of the words ὑμεῖς and φαγεῖν ("you" and "eat"), so that, despite their verbal agreement, there is a minor difference between Matthew and Mark on the one hand and Luke on the other. In fact, in subsequent editions of the Greek text, from NA^{26} onwards, the word order is the same in Luke as in Matthew and Mark, as it is in the great majority of manuscripts.

[34] It is interesting to note that John in his account of the feeding of the five thousand does retain the reference to the two hundred denarii at Mk 6:37b, omitted by Matthew and Luke, though putting it in the mouth of Philip (Jn 6:7). So this number too on the evidence of Mark and John would be a fixed point in the oral tradition, though not according to Matthew and Luke.

[35] France (2002), pp. 266-267

most of Mk 6:41, to a large extent word for word, Matthew and Luke both omit the last part of this verse, "and he divided the two fish among them all." They might have made these common omissions in Mk 6:39-41 independently, but it is much easier to envisage Matthew, say, doing so first and influencing Luke to do something similar.

Of the minor verbal agreements of Matthew and Luke against Mark, shown in the boxes in Figure 7.6, a particularly interesting one is the common insertion of ὡσεὶ ("about"), so that "five thousand men" becomes "about five thousand men". The use of ὡσεὶ, in the sense of "about", to qualify a number or a length of time or a distance, is characteristic of Luke. It appears in this sense not at all in Mark, once (in this pericope, at Mt 14:21) in Matthew, but 7 times in Luke and 4 times in Acts. So it is perhaps not surprising that Luke should insert it here, and in fact twice in Lk 9:14.[36] What is surprising is its insertion by Matthew, which might suggest that Matthew knew Luke and followed him here — a dependence which is in the opposite direction from what is more commonly proposed, as by Gundry and Goulder.

Another minor verbal agreement is of the words οἱ δὲ and οὐκ (literally, "but they" and "not") at Mt 14:17 || Mk 6:37 || Lk 9:13. As indicated in Neirynck,[37] the positive agreement of Matthew and Luke against Mark is stronger than shown by the exact verbal agreements. In Mark's account, it is Jesus who asks the disciples to find out how many loaves they have, but in Matthew and Luke, although phrased in different words, it is the disciples themselves who volunteer the information. That Matthew and Luke should independently make such a similar change to the storyline seems hard to believe, though the presence of an alternative oral tradition, known to both of them, could be used to provide an explanation. However, this is quite speculative, as we lack the knowledge of the historical background to assess how likely such circumstances are to have arisen.

7.5 The Great Commandment

Our next example is the Huck pericope number 208 with parallel texts Mt 22:34-40 || Mk 12:28-34 || Lk 10:25-28, which appears in Tables 3.8 and 6.12 as a candidate for further investigation. As opposed to our earlier examples of healing and feeding miracles, this pericope concerns the teaching of Jesus and, furthermore, has a number of complicating features that were not present in the earlier examples.

For one thing, assuming Markan priority, it appears that Luke has moved this pericope from its position in Mark to an earlier position, where it leads

[36] The second occurrence, though, is on the manuscript evidence doubtful.
[37] Neirynck (1974), p. 114

Table 7.5
Counts of verbal agreements with Mk 12:28-34

		Luke		
		0	1	total
Matthew	0	112 (98.86)	8 (19.52)	120
	1	13 (24.52)	20 (8.48)	33
	total	125	28	153

into Luke's well-known single tradition parable of the good Samaritan. Because in Luke the pericope occurs earlier in the sequence of pericopes than it does in Matthew and Mark, it appears twice in the Huck synopsis, as pericope number 208 ("The Great Commandment"), corresponding to its position in Matthew and Mark, and as pericope number 143 ("The Lawyer's Question"), corresponding to its position in Luke. This, incidentally, may serve as a reminder that differences in the relative positions of pericopes in the synoptic gospels, with discussion of the reasons for this, also feature in analysis of the synoptic problem, although it is not an aspect that is encompassed by our focus on verbal agreements. The parallel Greek texts are laid out in Figure 7.8 and the parallel English texts in Figure 7.9. Also included in Figures 7.8 and 7.9 are the verses Mt 22:46 and Lk 20:39-40. This is because Matthew has left the verse Mt 22:46 that parallels the comment in the last part of Mk 12:34, to the end of the next pericope, Huck number 209; and Luke, when moving the main body of the pericope to an earlier position in his gospel, has left behind the concluding verses, Lk 20:39-40, where they appear at the end of the previous pericope, Huck number 207, and provide a transition to pericope number 209. These verses from Matthew and Luke provide some parallel with the first part of Mk 12:32 and the last part of Mk 12:34.

There is a substantial piece of the text of Mark, in Mk 12:32b-34a, that is absent from both Matthew and Luke and so is unique to Mark. For the convenience of being able to fit the tables of parallel text on a single page, most of this passage, to the end of verse 33, has been omitted from Figures 7.8 and 7.9. Instead, the verses Mk 12:32-33, the response of the scribe to Jesus, are presented separately in both Greek and English in Figure 7.10.

In Table 7.5 we present the counts of verbal agreements aggregated over the 153 words in the section of text Mk 12:28-34 and in brackets the expected frequencies, given the row and column totals, under the hypothesis that Matthew and Luke are statistically independent in their verbal agreements with Mark. Compared with our earlier examples, the overall level of

Mt 22:34-40, 46	Mk 12:28-34	Lk 10:25-28, 20:39-40
34 Οἱ δὲ Φαρισαῖοι ἀκούσαντες ὅτι ἐφίμωσεν τοὺς Σαδδουκαίους συνήχθησαν ἐπὶ τὸ αὐτό, 35 καὶ ἐπηρώτησεν εἷς ἐξ αὐτῶν νομικὸς πειράζων αὐτόν· 36 διδάσκαλε, ποία ἐντολὴ μεγάλη ἐν τῷ νόμῳ; 37 ὁ δὲ ἔφη αὐτῷ· ἀγαπήσεις κύριον τὸν θεόν σου ἐν ὅλῃ τῇ καρδίᾳ σου καὶ ἐν ὅλῃ τῇ ψυχῇ σου καὶ ἐν ὅλῃ τῇ διανοίᾳ σου· 38 αὕτη ἐστὶν ἡ μεγάλη καὶ πρώτη ἐντολή. 39 δευτέρα ὁμοία αὐτῇ· ἀγαπήσεις τὸν πλησίον σου ὡς σεαυτόν. 40 ἐν ταύταις ταῖς δυσὶν ἐντολαῖς ὅλος ὁ νόμος κρέμαται καὶ οἱ προφῆται.	28 Καὶ προσελθὼν εἷς τῶν γραμματέων ἀκούσας αὐτῶν συζητούντων, εἰδὼς ὅτι καλῶς ἀπεκρίθη αὐτοῖς ἐπηρώτησεν αὐτόν· ποία ἐστὶν ἐντολὴ πρώτη πάντων; 29 ἀπεκρίθη ὁ Ἰησοῦς ὅτι πρώτη ἐστίν· ἄκουε, Ἰσραήλ, κύριος ὁ θεὸς ἡμῶν κύριος εἷς ἐστιν, 30 καὶ ἀγαπήσεις κύριον τὸν θεόν σου ἐξ ὅλης τῆς καρδίας σου καὶ ἐξ ὅλης τῆς ψυχῆς σου καὶ ἐξ ὅλης τῆς διανοίας σου καὶ ἐξ ὅλης τῆς ἰσχύος σου. 31 δευτέρα αὕτη· ἀγαπήσεις τὸν πλησίον σου ὡς σεαυτόν. μείζων τούτων ἄλλη ἐντολὴ οὐκ ἔστιν.	25 Καὶ ἰδοὺ νομικός τις ἀνέστη ἐκπειράζων αὐτὸν λέγων· διδάσκαλε, τί ποιήσας ζωὴν αἰώνιον κληρονομήσω; 26 ὁ δὲ εἶπεν πρὸς αὐτόν· ἐν τῷ νόμῳ τί γέγραπται; πῶς ἀναγινώσκεις; 27 ὁ δὲ ἀποκριθεὶς εἶπεν· ἀγαπήσεις κύριον τὸν θεόν σου ἐξ ὅλης τῆς καρδίας σου καὶ ἐν ὅλῃ τῇ ψυχῇ σου καὶ ἐν ὅλῃ τῇ ἰσχύϊ σου καὶ ἐν ὅλῃ τῇ διανοίᾳ σου, καὶ τὸν πλησίον σου ὡς σεαυτόν. 28 εἶπεν δὲ αὐτῷ· ὀρθῶς ἀπεκρίθης· τοῦτο ποίει καὶ ζήσῃ.
	32 Καὶ εἶπεν αὐτῷ ὁ γραμματεύς· καλῶς, διδάσκαλε . . .	20:39 Ἀποκριθέντες δέ τινες τῶν γραμματέων εἶπαν· διδάσκαλε, καλῶς εἶπας.
46 καὶ οὐδεὶς ἐδύνατο ἀποκριθῆναι αὐτῷ λόγον οὐδὲ ἐτόλμησέν τις ἀπ᾽ ἐκείνης τῆς ἡμέρας ἐπερωτῆσαι αὐτὸν οὐκέτι.	34 καὶ ὁ Ἰησοῦς ἰδὼν αὐτὸν ὅτι νουνεχῶς ἀπεκρίθη εἶπεν αὐτῷ· οὐ μακρὰν εἶ ἀπὸ τῆς βασιλείας τοῦ θεοῦ. Καὶ οὐδεὶς οὐκέτι ἐτόλμα αὐτὸν ἐπερωτῆσαι.	20:40 οὐκέτι γὰρ ἐτόλμων ἐπερωτᾶν αὐτὸν οὐδέν.

Figure 7.8
The Great Commandment (NA[25] Greek)

Mt 22:34-40, 46	Mk 12:28-34	Lk 10:25-28, 20:39-40
34 When the Pharisees heard that he had silenced the Sadducees, they gathered together, 35 and one of them, a lawyer, asked him a question to test him. 36 "Teacher, which commandment in the law is the greatest?" 37 He said to him, " 'You shall love the Lord your God with all your heart, and with all your soul, and with all your mind.' 38 This is the greatest and first commandment. 39 And a second is like it: 'You shall love your neighbour as yourself.' 40 On these two commandments hang all the law and the prophets."	28 One of the scribes came near and heard them disputing with one another, and seeing that he answered them well, he asked him, "Which commandment is the first of all?" 29 Jesus answered, "The first is, 'Hear, O Israel: the Lord our God, the Lord is one; 30 you shall love the Lord your God with all your heart, and with all your soul, and with all your mind, and with all your strength.' 31 The second is this, 'You shall love your neighbour as yourself.' There is no other commandment greater than these." 32 Then the scribe said to him, "You are right, Teacher; . . .	25 Just then a lawyer stood up to test Jesus. Teacher, he said, what must I do to inherit eternal life? 26 He said to him, What is written in the law? What do you read there? 27 He answered, You shall love the Lord your God with all your heart, and with all your soul, and with all your strength, and with all your mind; and your neighbour as yourself. 28 And he said to him, You have given the right answer; do this, and you will live.
		20:39 Then some of the scribes answered, Teacher, you have spoken well.
46 No one was able to give him an answer, nor from that day did anyone dare to ask him any more questions.	34 When Jesus saw that he answered wisely, he said to him, "You are not far from the kingdom of God." After that no one dared to ask him any question.	20:40 For they no longer dared to ask him another question.

Figure 7.9
The Great Commandment (NRSV translation)

Greek (NA^{25})	English (NRSV)
32 Καὶ εἶπεν αὐτῷ ὁ γραμματεύς· καλῶς, διδάσκαλε, ἐπ᾽ ἀληθείας εἶπες ὅτι εἷς ἐστιν καὶ οὐκ ἔστιν ἄλλος πλὴν αὐτοῦ· 33 καὶ τὸ ἀγαπᾶν αὐτὸν ἐξ ὅλης τῆς καρδίας καὶ ἐξ ὅλης τῆς συνέσεως καὶ ἐξ ὅλης τῆς ἰσχύος καὶ τὸ ἀγαπᾶν τὸν πλησίον ὡς ἑαυτὸν περισσότερόν ἐστιν πάντων τῶν ὁλοκαυτωμάτων καὶ θυσιῶν.	32 Then the scribe said to him, "You are right, Teacher; you have truly said that 'he is one, and besides him there is no other'; 33 and 'to love him with all the heart, and with all the understanding, and with all the strength,' and 'to love one's neighbour as oneself,' — this is much more important than all whole burnt offerings and sacrifices."

Figure 7.10
The scribe's response (Mk 12:32-33)

verbal agreements of Matthew and Luke with Mark is considerably lower, though this is largely attributable to the common omission by Matthew and Luke of most of Mk 12:32-34a. In any case, yet again, as in our previous examples, we see that the observed frequencies along the diagonal of the table are much greater than the expected frequencies, which strongly indicates a positive association between the texts of Matthew and Luke in what words are retained unchanged from Mark. We may also observe in Figure 7.8 the 14 boxed words that are minor verbal agreements of Matthew and Luke against Mark.

A feature of the present pericope, not seen in our previous examples, is that, as reported by the gospel authors, Jesus quotes biblical texts, specifically in this case from the Book of Deuteronomy, Deut 6:4-5, and from the Book of Leviticus, Lev 19:18. The quotations are originally from Hebrew scripture, but the gospel authors, and in particular Mark as the first, writing in Greek, are likely to be using or at the very least to be influenced by some version of the translation from the Hebrew into Greek, known as the *Septuagint* and commonly referred to by the abbreviation LXX. This was the form in which the scriptures were known to the Hellenistic Jews of the diaspora around the Mediterranean and to the early Greek-speaking Church.[38] In Figures 7.11 and 7.12 we present the Greek LXX text of Deut 6:4b-5 and another relevant text from the Second Book of Kings, 2Ki 23:25, together with the NETS

[38]The modern critical edition of the Septuagint that we have used for reference is that of Rahlfs (1979). A recent translation of the Septuagint into English ("A New English Translation of the Septuagint", NETS) may be found in Pietersma and Wright (2007). For an introduction to the Septuagint in historical context see Dines (2004), Law (2013) or, with special reference to the *Codex Sinaiticus*, Parker (2010). See Law (2013), pp. 99f., for discussion of the use of the Septuagint in the gospels.

Greek (LXX)	English (NETS)
4 Ἄκουε, Ἰσραήλ· κύριος ὁ θεὸς ἡμῶν κύριος εἷς ἐστιν· 5 καὶ ἀγαπήσεις κύριον τὸν θεόν σου ἐξ ὅλης τῆς καρδίας σου καὶ ἐξ ὅλης τῆς ψυχῆς σου καὶ ἐξ ὅλης τῆς δυνάμεώς σου.	4 Hear, O Israel: The Lord our God is one Lord. 5 And you shall love the Lord your God with the whole of your mind and with the whole of your soul and with the whole of your power.

Figure 7.11
The Shema (Deut 6:4b-5)

Greek (LXX)	English (NETS)
ὅμοιος αὐτῷ οὐκ ἐγενήθη ἔμπροσθεν αὐτοῦ βασιλεύς, ὃς ἐπέστρεψεν πρὸς κύριον ἐν ὅλῃ καρδίᾳ αὐτοῦ καὶ ἐν ὅλῃ ψυχῇ αὐτοῦ καὶ ἐν ὅλῃ ἰσχύι αὐτοῦ κατὰ πάντα τὸν νόμον Μωυσῆ, καὶ μετ᾽ αὐτὸν οὐκ ανέστη ὅμοιος αὐτῷ.	Before him there was no king like him, who turned to the Lord with his whole heart and with his whole soul and with his whole strength, according to all the law of Moyses, and after him none arose like him.

Figure 7.12
King Josiah (2Ki 23:25)

translations into English.[39] These will be useful when we come to discuss the synoptic parallels of Figure 7.8.

As Davies and Allison[40] readily acknowledge, this is a pericope in which the numerous and significant agreements of Matthew and Luke against Mark provide a serious challenge to the two-source hypothesis.They review a number of possible explanations but tentatively suggest their preferred option that Matthew and Luke knew the pericope as it stood in Mark and also through oral tradition.

At the centre of this pericope stands the great commandment from the Shema (LXX Deut 6:4b-5, Figure 7.11), which is at the heart of Jewish religious life, in combination with the commandment from Lev 19:18 to "love your neighbour as yourself." In Mark it is Jesus who states these commandments, which are then repeated in his own words by the scribe to whose question Jesus was responding (Mk 12:32-33, Figure 7.10); in Matthew they are stated by Jesus; but in Luke Jesus prompts the lawyer who is testing him to state them himself. Despite the differences, there is thus a core of common material

[39]For the NETS translation in Figure 7.11 the alternative LXX reading διανοίας instead of καρδίας is being used to give the translation "mind" instead of "heart". See also the note in Table 7.6.

[40]Davies and Allison (1997), Volume III, pp. 235-236

Table 7.6
Words used in the Shema

LXX Deut 6:5	LXX 2Ki 23:25	Mk 12:30	Mk 12:33	Mt 22:37	Lk 10:27
καρδία	καρδία	καρδία	καρδία	καρδία	καρδία
ψυχή	ψυχή	ψυχή	σύνεσις	ψυχή	ψυχή
δύναμις	ἰσχύς	διάνοια	ἰσχύς	διάνοια	ἰσχύς
		ἰσχύς			διάνοια
Note: there is a significant variant reading in the *Codex Vaticanus* text of LXX Deut 6:5, διάνοια for καρδία					
ἐξ	ἐν	ἐξ	ἐξ	ἐν	ἐξ, ἐν

that is present in all three of the synoptic gospels. However, we may note two major common omissions by Matthew and Luke. Firstly, they both omit the first verse (Deut 6:4) of the Shema, "Hear, O Israel, The Lord our God is one Lord", that is present in Mk 12:29. Secondly, they both omit the repetition of the commandments by the scribe at Mk 12:32-33.

At a more detailed level, there are noteworthy differences between the gospels themselves and between them and the LXX text of Deut 6:5 as we now have it. In the English text of Mt 22:37 || Mk 12:30 || Lk 10:27 in Figure 7.9 the differences are fewer, but this masks some interesting and significant differences in the underlying Greek text of Figure 7.8, which are summarised in Table 7.6. In Mk 12:30, there are four phrases to describe the way in which God is to be loved, "with all your heart (καρδία), and with all your soul (ψυχή), and with all your mind (διάνοια), and with all your strength (ἰσχύς)." In LXX Deut 6:5, as shown in Figure 7.11, there are only three phrases. Mark has the three phrases of Deut 6:5, but with δύναμις (power, might, strength) replaced by the synonymous ἰσχύς, which also appears in the quotation from the Shema in LXX 2Ki 23:25, as shown in Figure 7.12. Furthermore, Mark in the third position adds the fourth phrase with διάνοια, which appears in place of καρδία in the *Codex Vaticanus* text of LXX Deut 6:5. In the scribe's version in Mk 12:33, as given in Figure 7.10, there are three phrases, with σύνεσις (understanding) replacing ψυχή. Mt 22:37 has three phrases, similar to the first three phrases of Mark 12:30, but dropping the fourth phrase. Lk 10:27 has four phrases, similar to those of Mk 12:30, but interchanging the third and fourth. Another type of difference between the texts arises from the fact that behind the word "with" in the NRSV translation there lies either the preposition ἐξ, which takes the genitive case, or the preposition ἐν, which takes the dative. Mark follows LXX Deut 6:5 in using ἐξ, but Matthew uses ἐν, which is used in LXX 2Ki 23:25. However, Luke uses ἐξ in the first phrase but then uses ἐν in the remaining three.

How are we to account for this complex variation in the text quoted from the Shema? Firstly, we should be aware that there will have been considerable textual plurality, a number of variants of the LXX and possibly other Greek texts of the scriptures in circulation;[41] but according to France, with regard to the present case:[42] "These variations indicate a text in regular liturgical or catechetical use." We may suppose that in the early church the Shema text of Deut 6:5 was in common and regular use, as it was in the Jewish community of which much of the church was still a part or with which it still had some interaction, and that variant versions had arisen, perhaps through the interplay of oral transmission with written materials, which did not affect the essential meaning. Assuming that Matthew had Mark as a source, he may still have decided to use the Shema in a form that was familiar to him, in particular reverting to the use of three phrases, as in our LXX text, instead of Mark's four. More puzzling is Luke's version, in particular his switch from the use of ἐξ in the first phrase to ἐν in the remaining three. For Gundry,[43] this suggests a conflation of Mark and Matthew. In a similar vein, to Goulder,[44] who also notes that Luke's reversal of order of the last two phrases in Mark brings διάνοια to the end, as in Matthew, "it seems a clear instance of mixed citation from memory."

One of the major differences between Mark on the one hand and Matthew and Luke on the other with regard to the present pericope is that in Mark the scribe who questions Jesus appears to do so in an open, honest manner, as a seeker after truth, but in Matthew and Luke the lawyer who questions Jesus appears to be hostile, as one who seeks to test him. In Mk 12:32-33 the scribe is shown as approving of and elaborating upon Jesus' answer, and at Mk 12:34 Jesus in turn shows his approval of the scribe's response. The different scenario in Matthew and Luke is brought out, at least to some extent, in some of their minor agreements against Mark, to which we now turn.

Both Matthew and Luke have νομικὸς ("lawyer") instead of Mark's γραμματεύς ("scribe") and they both insert the phrase ἐν τῷ νόμῳ ("in the law"), which is absent from Mark. The attention drawn in this way to the law and its interpretation perhaps in itself carries some suggestion of polemic. In addition Matthew and Luke have the common insertion that Jesus is addressed as διδάσκαλε ("teacher"). All these minor verbal agreements are shown as boxed material in Figure 7.8, but more indicative of the change in atmosphere in Matthew and Luke is the presence of the words πειράζων (a participle, "testing", translated "to test") at Mt 22:35 and ἐκπειράζων at Lk 10:25. This does not appear as a minor verbal agreement in Figure 7.8, nor is it counted as

[41]For background, see Dines (2004), Chapter 5, and, more specifically, Law (2013), p. 86. For a more detailed listing of variants, Wevers (1977) may be consulted.

[42]France (2002), p. 480

[43]Gundry (1994), p. 449

[44]Goulder (1989), p. 485

such in our statistics, because the words are not exactly identical, but the addition of the prefix ἐκ does not change the meaning of the verb πειράζω. The two words are translated identically, so that they do appear boxed as a minor verbal agreement in the English text of Figure 7.9. They are also recorded as a minor agreement in Neirynck.[45] The significance of the verb πειράζω may be gauged by noting that it appears, translated as "tempt", in all three synoptic gospels in the account of Jesus' temptation by the devil in the wilderness, as does the verb ἐκπειράζω, quoting LXX Deut 6:16, in Matthew's and Luke's accounts (Mt 4:1-11 || Mk 1:12-13 || Lk 4:1-13).

Overall, the agreements of various kinds of Matthew and Luke against Mark are so many that it seems inconceivable that they were working independently of each other, unless they had some other common source, either written or oral. However, the editors of *The Critical Edition of Q* decided that the material shared by Matthew and Luke in this pericope was not to be assigned to the Q source,[46] and the suggestion of Davies and Allison, referred to above, of an oral tradition known to Matthew and Luke, like all such appeals to the influence of oral tradition, leads to the question of how plausible it is that Matthew and Luke had access to oral traditions that must have been very similar, yet apparently unknown to Mark or at least not used by him. A hypothesis that Luke had both Mark and Matthew as a source makes it much easier to explain the agreements of Matthew and Luke as well as the puzzle of the variant versions of the citation from the Shema.

7.6 About David's Son

Our next example is the short Huck pericope number 209, with parallel texts Mt 22:41-46 || Mk 12:35-37a || Lk 20:41-44, which, in Mark and Matthew, follows on immediately from the pericope of The Great Commandment studied in the previous section. The present pericope appears in Table 3.8 as a candidate for further investigation, and sections of text that include all or part of it appear in Table 5.6 (Mk 12:36-38) and in Table 5.7 (Mk 12:35-44).

In Figures 7.13 and 7.14 we give the parallel texts of Mt 22:41-45 || Mk 12:35-37a || Lk 20:41-44, in Greek and English, respectively, omitting Mt 22:46, which, as we saw in the previous section, has its Markan parallel in pericope number 208.

In Table 7.7 we present the counts of verbal agreements aggregated over the 56 words in the section of text Mk 12:35-37a and in brackets the expected frequencies, given the row and column totals, under the hypothesis that Matthew and Luke are statistically independent in their verbal agree-

[45] Neirynck (1974), p. 157
[46] Robinson et al. (2000), pp. 200-205

Mt 22:41-45	Mk 12:35-37a	Lk 20:41-44
41 Συνηγμένων δὲ τῶν Φαρισαίων ἐπηρώτησεν αὐτοὺς ὁ Ἰησοῦς 42 λέγων· τί ὑμῖν δοκεῖ περὶ τοῦ χριστοῦ; τίνος υἱός ἐστιν; λέγουσιν αὐτῷ· τοῦ Δαυίδ. 43 λέγει αὐτοῖς· πῶς οὖν Δαυὶδ ἐν πνεύματι καλεῖ αὐτὸν κύριον λέγων·	35 Καὶ ἀποκριθεὶς ὁ Ἰησοῦς ἔλεγεν διδάσκων ἐν τῷ ἱερῷ· πῶς λέγουσιν οἱ γραμματεῖς ὅτι ὁ χριστὸς υἱὸς Δαυίδ ἐστιν; 36 αὐτὸς Δαυὶδ εἶπεν ἐν τῷ πνεύματι τῷ ἁγίῳ·	41 Εἶπεν δὲ πρὸς αὐτούς· πῶς λέγουσιν τὸν χριστὸν εἶναι Δαυὶδ υἱόν; 42 αὐτὸς γὰρ Δαυὶδ λέγει ἐν βίβλῳ ψαλμῶν·
44 εἶπεν κύριος τῷ κυρίῳ μου· κάθου ἐκ δεξιῶν μου, ἕως ἂν θῶ τοὺς ἐχθρούς σου ὑποκάτω τῶν ποδῶν σου;	εἶπεν κύριος τῷ κυρίῳ μου· κάθου ἐκ δεξιῶν μου, ἕως ἂν θῶ τοὺς ἐχθρούς σου ὑποκάτω τῶν ποδῶν σου;	εἶπεν κύριος τῷ κυρίῳ μου· κάθου ἐκ δεξιῶν μου, 43 ἕως ἂν θῶ τοὺς ἐχθρούς σου ὑποπόδιον τῶν ποδῶν σου.
45 εἰ οὖν Δαυὶδ καλεῖ αὐτὸν κύριον, πῶς υἱὸς αὐτοῦ ἐστιν;	37 αὐτὸς Δαυὶδ λέγει αὐτὸν κύριον, καὶ πόθεν αὐτοῦ ἐστιν υἱός;	44 Δαυὶδ οὖν αὐτὸν κύριον καλεῖ, καὶ πῶς αὐτοῦ υἱός ἐστιν;

Figure 7.13
About David's Son (NA[25] Greek)

Mt 22:41-45	Mk 12:35-37a	Lk 20:41-44
41 Now while the Pharisees were gathered together, Jesus asked them this question: 42 "What do you think of the Messiah? Whose son is he?" They said to him, "The son of David." 43 He said to them, "How is it then that David by the Spirit calls him Lord, saying,	35 While Jesus was teaching in the temple, he said, "How can the scribes say that the Messiah is the son of David? 36 David himself, by the Holy Spirit, declared,	41 Then he said to them, "How can they say that the Messiah is David's son? 42 For David himself says in the book of Psalms,
44 'The Lord said to my Lord, "Sit at my right hand, until I put your enemies under your feet." '?	'The Lord said to my Lord, "Sit at my right hand, until I put your enemies under your feet." '?	'The Lord said to my Lord, "Sit at my right hand, 43 until I make your enemies your footstool." '
45 If David thus calls him Lord, how can he be his son?"	37 David himself calls him Lord; so how can he be his son?"	44 David thus calls him Lord; so how can he be his son?"

Figure 7.14
About David's Son (NRSV translation)

Table 7.7
Counts of verbal agreements with Mk 12:35-37a

		Luke		
		0	1	total
Matthew	0	19 (10.27)	4 (12.73)	23
	1	6 (14.73)	27 (18.27)	33
	total	25	31	56

ments with Mark. Yet again, as in earlier examples, the observed frequencies along the diagonal of the table are substantially greater than the expected frequencies, which indicates a positive association between the texts of Matthew and Luke in what words are retained unchanged from Mark, though here, because of the smaller number of words than in earlier passages, the evidence from the frequencies appears perhaps somewhat less compelling. A feature worth noting here is that the relatively high proportion of verbal agreements overall is due to the presence of a biblical quotation at Mt 22:44 || Mk 12:36b || Lk 20:42b-43, which is reproduced in almost exactly the same words in each gospel. We may also observe in Figure 7.13 the 5 boxed words that are minor verbal agreements of Matthew and Luke against Mark.

In the first part of this pericope, Mt 22:41-43 || Mk 12:35-36a || Lk 20:41-42a, both Matthew and Luke have freely adapted Mark's introduction to create their own scenarios, or have drawn on different oral traditions, and there is much variation in the wording. In Mark and Luke, Jesus poses a question to his listeners about what, in the case of Mark, the scribes say and, in the case of Luke, the unspecified "they" say. In Matthew the monologue is replaced by a dialogue, in which Jesus asks the pharisees to say what they think. After this introduction, however, all three gospels parallel each other very closely in presenting the paradox posed by Jesus about the nature of his lordship as Messiah.

A central feature of the pericope is that Jesus quotes a biblical text much quoted or alluded to in early Christian literature, Psalm 110:1 (LXX Psalm 109:1), which in the LXX, with the NETS translation, reads

> Εἶπεν ὁ κύριος τῷ κυρίῳ μου Κάθου ἐκ δεξιῶν μου, ἕως ἂν θῶ τοὺς ἐχθρούς σου ὑποπόδιον τῶν ποδῶν σου.
>
> The Lord said to my lord, "Sit on my right until I make your enemies a footstool for your feet."

Mark has omitted the article ὁ before κύριος ("Lord") and altered ὑποπόδιον ("footstool") to the preposition ὑποκάτω ("under"). In so doing, Mark may be influenced by a phrase from Psalm 8:6 (LXX Psalm 8:7),

> πάντα ὑπέταξας ὑποκάτω τῶν ποδῶν αὐτοῦ
>
> you subjected all under his feet.

Mark is followed word for word by Matthew. Luke also omits the article but reverts to ὑποπόδιον, as in LXX Psalm 109:1. So here we have a situation where there is a substantial common retention by Matthew and Luke, apart from the word ὑποκάτω. Now it may well be expected that scriptural quotations should lead to word-for-word agreements, so this common retention does not provide a very strong argument for dependence between Matthew and Luke. However, it is not always the case that scriptural quotations are faithfully reproduced from a common source, as shown by the example of the citation from the Shema in the previous section, and it is noteworthy that both Matthew and Luke follow Mark in omitting the article ὁ from the LXX as we have it.

A much stronger case for dependence may be found in the final verse of this pericope, Mt 22:45 || Mk 12:37a || Lk 20:44. Here Jesus challenges his audience with a question, where there is substantial common retention, a common omission, and minor verbal agreements of Matthew and Luke against Mark. The common retentions are

> Δαυὶδ ... αὐτὸν κύριον ... αὐτοῦ ἐστιν υἱός
>
> David ... him Lord ... is his son

although it is worth noting that the last three words in the Greek occur in a different order in each of the gospels. The common omission is of the word αὐτὸς ("himself"), Matthew and Luke both insert the word οὖν ("thus"), and they both replace λέγει by καλεῖ ("calls") and πόθεν by πῶς ("how"). These two replacements, insignificant as far as the meaning of the text is concerned, do not affect the English translation, so that they are not apparent there. It seems implausible that Matthew and Luke would independently have made exactly these same alterations, *pace* Davies and Allison,[47] who with regard

[47] Davies and Allison (1997), Volume III, p. 249

to this pericope assert that "the agreements ... against Mark are minor and due to independent editing." The fact that these agreements of Matthew and Luke against Mark are minor in terms of their impact on the meaning of the text does not make them any less difficult to explain under the assumption that Matthew and Luke were using the text of Mark independently of each other.

7.7 Jesus Mocked and Peter's Triple Denial

In Table 6.12, the lengthy Huck pericope number 241, Mt 26:57-75 || Mk 14:53-72 || Lk 22:54-71, was found to be one of the pericopes with the largest proportions of minor verbal agreements of Matthew and Luke against Mark, and in Table 5.7 a section of this pericope, Mk 14:65-69, emerged as of potential interest with regard to the dependence of Luke and Matthew in their use of Mark. In terms of the underlying HMM, this section of text consists essentially of a long decoded State 2 passage, with the corresponding model Equation (5.17) in Section 5.2. The fitted regression coefficients in this equation are such as to indicate that generally, in State 2, Luke was following Mark very loosely, suggestive of the influence of oral tradition, but also with a strong positive association between the words in Mark that were retained by Luke and those that were retained by Matthew, suggestive of the influence of Matthew on Luke, or vice versa.

In terms of the gospel narrative, after our previous examples of healing, miracle working and the teaching of Jesus, the present pericope falls within the Passion Narrative and deals with the events following the arrest of Jesus: the examination of Jesus by the Jewish Council, the Sanhedrin, and the mocking of Jesus, but interwoven with the story of Peter's triple denial of Jesus. It is a complicated pericope with a complex web of similarities and dissimilarities between the three synoptic parallels. One major difference between the three accounts is that, whereas in Matthew and Mark the examination, condemnation and mocking of Jesus come before the account of Peter's denials, all during the course of the same night, Luke deals with Peter's denials first, then with the mocking of Jesus, and only after that with the trial, which in Luke occurs the following morning. Upon more detailed examination of the texts of Matthew, Mark and Luke and of the sequence of hidden states given by the decoding described in Section 5.3, it turned out to be fruitful to consider a slightly longer section of text than the one indicated in Table 5.7, so as to cover not only the mocking of Jesus but also the whole of the account of Peter's denials. We present the parallel texts in two pairs of figures, the Greek and English texts of the mocking of Jesus, Mt 26:67-68 || Mk 14:65 || Lk 22:63-65, in Figures 7.15 and 7.16, respectively, and the Greek and English texts of Peter's denials, Mt 26:69-75 || Mk 14:66-72 || Lk 22:55b-62, in

Table 7.8
Counts of verbal agreements with Mk 14:65-72

		Luke		
		0	1	total
Matthew	0	72 (59.85)	14 (26.15)	86
	1	31 (43.15)	31 (18.85)	62
	total	103	45	148

Figures 7.17 and 7.18, respectively. In the Greek text of Mark in Figure 7.17, we include in square brackets at the top the three words πρὸς τὸ φῶς ("at the fire" or "in the firelight") from Mk 14:54, to show the source of the verbal agreement with Mark that occurs in Lk 22:56.

A textual issue worth noting is that at the end of Mk 14:68, in Figures 7.17 and 7.18, we include in double square brackets the words καὶ ἀλέκτωρ ἐφώνησεν ("Then the cock crowed"), which are present in NA^{28}, but not in NA^{25}, and so are not included in our statistics. The textual evidence is fairly evenly divided as to whether these words should be included as original to Mark's gospel, but the matter is of some importance for the synoptic problem, as we shall discuss later.

In Table 7.8 we present the counts of verbal agreements aggregated over the 148 words in the text of Mk 14:65-72, as presented in Figures 7.15 and 7.17, and in brackets the expected frequencies, given the row and column totals, under the hypothesis that Matthew and Luke are statistically independent in their verbal agreements with Mark. The observed frequencies along the diagonal of the table are greater than the expected frequencies. Yet again there appears to be a positive association between Matthew and Luke in their verbal agreements with Mark. We may also observe in Figures 7.15 and 7.17 the 20 boxed words that are minor verbal agreements of Matthew and Luke against Mark, which turn out to be of particular importance in this example.

Although the essentials of the story are the same in all three synoptic accounts, there is a great deal of variation in the details, and this is one of the pericopes used by Dunn (2003a) to illustrate the influence of oral tradition.[48] We may envisage that the traumatic events leading up to the crucifixion, seared especially into the memory of Peter, would have been recounted again and again, first by Peter and then by others, which could have led to a multiplicity of variant forms in the presentation.

[48] Dunn (2003a), pp. 168-169

Mt 26:67-68	Mk 14:65	Lk 22:63-65
67 Τότε ἐνέπτυσαν εἰς τὸ πρόσωπον αὐτοῦ καὶ ἐκολάφισαν αὐτόν, οἱ δὲ ἐράπισαν 68 λέγοντες· προφήτευσον ἡμῖν, χριστέ, τίς ἐστιν ὁ παίσας σε;	65 Καὶ ἤρξαντό τινες ἐμπτύειν αὐτῷ καὶ περικαλύπτειν αὐτοῦ τὸ πρόσωπον καὶ κολαφίζειν αὐτὸν καὶ λέγειν αὐτῷ· προφήτευσον, καὶ οἱ ὑπηρέται ῥαπίσμασιν αὐτὸν ἔλαβον.	63 Καὶ οἱ ἄνδρες οἱ συνέχοντες αὐτὸν ἐνέπαιζον αὐτῷ δέροντες, 64 καὶ περικαλύψαντες αὐτὸν ἐπηρώτων λέγοντες· προφήτευσον, τίς ἐστιν ὁ παίσας σε; 65 καὶ ἕτερα πολλὰ βλασφημοῦντες ἔλεγον εἰς αὐτόν.

Figure 7.15
Jesus Mocked (NA^{25} Greek)

Mt 26:67-68	Mk 14:65	Lk 22:63-65
67 Then they spat in his face and struck him; and some slapped him, 68 saying, "Prophesy to us, you Messiah! Who is it that struck you?"	65 Some began to spit on him, to blindfold him, and to strike him, saying to him, "Prophesy!" The guards also took him over and beat him.	63 Now the men who were holding Jesus began to mock him and beat him; 64 they also blindfolded him and kept asking him, "Prophesy! Who is it that struck you?" 65 They kept heaping many other insults on him.

Figure 7.16
Jesus Mocked (NRSV translation)

Mt 26:69-75	Mk 14:66-72	Lk 22:55b-62
69 Ὁ δὲ Πέτρος ἐκάθητο ἔξω ἐν τῇ αὐλῇ· καὶ προσῆλθεν αὐτῷ μία παιδίσκη λέγουσα· καὶ σὺ ἦσθα μετὰ Ἰησοῦ τοῦ Γαλιλαίου. 70 ὁ δὲ ἠρνήσατο ἔμπροσθεν πάντων λέγων· οὐκ οἶδα τί λέγεις. 71 ἐξελθόντα δὲ εἰς τὸν πυλῶνα εἶδεν αὐτὸν ἄλλη καὶ λέγει τοῖς ἐκεῖ· οὗτος ἦν μετὰ Ἰησοῦ τοῦ Ναζωραίου. 72 καὶ πάλιν ἠρνήσατο μετὰ ὅρκου ὅτι οὐκ οἶδα τὸν ἄνθρωπον. 73 μετὰ μικρὸν δὲ προσελθόντες οἱ ἑστῶτες εἶπον τῷ Πέτρῳ· ἀληθῶς καὶ σὺ ἐξ αὐτῶν εἶ, καὶ γὰρ ἡ λαλιά σου δῆλόν σε ποιεῖ. 74 τότε ἤρξατο καταθεματίζειν καὶ ὀμνύειν ὅτι οὐκ οἶδα τὸν ἄνθρωπον. καὶ εὐθὺς ἀλέκτωρ ἐφώνησεν. 75 καὶ ἐμνήσθη ὁ Πέτρος τοῦ ῥήματος Ἰησοῦ εἰρηκότος ὅτι πρὶν ἀλέκτορα φωνῆσαι τρὶς ἀπαρνήσῃ με· καὶ ἐξελθὼν ἔξω ἔκλαυσεν πικρῶς.	[54 ... πρὸς τὸ φῶς.] 66 Καὶ ὄντος τοῦ Πέτρου κάτω ἐν τῇ αὐλῇ ἔρχεται μία τῶν παιδισκῶν τοῦ ἀρχιερέως 67 καὶ ἰδοῦσα τὸν Πέτρον θερμαινόμενον ἐμβλέψασα αὐτῷ λέγει· καὶ σὺ μετὰ τοῦ Ναζαρηνοῦ ἦσθα τοῦ Ἰησοῦ. 68 ὁ δὲ ἠρνήσατο λέγων· οὔτε οἶδα οὔτε ἐπίσταμαι σὺ τί λέγεις. καὶ ἐξῆλθεν ἔξω εἰς τὸ προαύλιον [[καὶ ἀλέκτωρ ἐφώνησεν]]. 69 καὶ ἡ παιδίσκη ἰδοῦσα αὐτὸν ἤρξατο πάλιν λέγειν τοῖς παρεστῶσιν ὅτι οὗτος ἐξ αὐτῶν ἐστιν. 70 ὁ δὲ πάλιν ἠρνεῖτο. καὶ μετὰ μικρὸν πάλιν οἱ παρεστῶτες ἔλεγον τῷ Πέτρῳ· ἀληθῶς ἐξ αὐτῶν εἶ, καὶ γὰρ Γαλιλαῖος εἶ. 71 ὁ δὲ ἤρξατο ἀναθεματίζειν καὶ ὀμνύναι ὅτι οὐκ οἶδα τὸν ἄνθρωπον τοῦτον ὃν λέγετε. 72 καὶ εὐθὺς ἐκ δευτέρου ἀλέκτωρ ἐφώνησεν. καὶ ἀνεμνήσθη ὁ Πέτρος τὸ ῥῆμα ὡς εἶπεν αὐτῷ ὁ Ἰησοῦς ὅτι πρὶν ἀλέκτορα δὶς φωνῆσαι τρίς με ἀπαρνήσῃ· καὶ ἐπιβαλὼν ἔκλαιεν.	55 ... ἐν ... ἐκάθητο ὁ Πέτρος μέσος αὐτῶν. 56 ἰδοῦσα δὲ αὐτὸν παιδίσκη τις καθήμενον πρὸς τὸ φῶς καὶ ἀτενίσασα αὐτῷ εἶπεν· καὶ οὗτος σὺν αὐτῷ ἦν. 57 ὁ δὲ ἠρνήσατο λέγων· οὐκ οἶδα αὐτόν, γύναι. 58 καὶ μετὰ βραχὺ ἕτερος ἰδὼν αὐτὸν ἔφη· καὶ σὺ ἐξ αὐτῶν εἶ. ὁ δὲ Πέτρος ἔφη· ἄνθρωπε, οὐκ εἰμί. 59 καὶ διαστάσης ὡσεὶ ὥρας μιᾶς ἄλλος τις διϊσχυρίζετο λέγων· ἐπ᾽ ἀληθείας καὶ οὗτος μετ᾽ αὐτοῦ ἦν, καὶ γὰρ Γαλιλαῖός ἐστιν. 60 εἶπεν δὲ ὁ Πέτρος· ἄνθρωπε, οὐκ οἶδα ὃ λέγεις. καὶ παραχρῆμα ἔτι λαλοῦντος αὐτοῦ ἐφώνησεν ἀλέκτωρ. 61 καὶ στραφεὶς ὁ κύριος ἐνέβλεψεν τῷ Πέτρῳ, καὶ ὑπεμνήσθη ὁ Πέτρος τοῦ λόγου τοῦ κυρίου ὡς εἶπεν αὐτῷ ὅτι πρὶν ἀλέκτορα φωνῆσαι σήμερον ἀπαρνήσῃ με τρίς. 62 καὶ ἐξελθὼν ἔξω ἔκλαυσεν πικρῶς.

Figure 7.17
Peter's Denial (NA^{25} Greek)

Mt 26:69-75

69 Now Peter was sitting outside in the courtyard. A servant-girl came to him and said, "You also were with Jesus the Galilean." 70 But he denied it before all of them, saying, "I do not know what you are talking about." 71 When he went out to the porch, another servant-girl saw him, and she said to the bystanders, "This man was with Jesus of Nazareth." 72 Again he denied it with an oath, "I do not know the man." 73 After a little while the bystanders came up and said to Peter, "Certainly you are also one of them, for your accent betrays you." 74 Then he began to curse, and he swore an oath, "I do not know the man!" At that moment the cock crowed. 75 Then Peter remembered what Jesus had said: "Before the cock crows, you will deny me three times." And he went out and wept bitterly.

Mk 14:66-72

66 While Peter was below in the courtyard, one of the servant-girls of the high priest came by. 67 When she saw Peter warming himself, she stared at him and said, "You also were with Jesus, the man from Nazareth." 68 But he denied it, saying, "I do not know or understand what you are talking about." And he went out into the forecourt. [[Then the cock crowed.]] 69 And the servant-girl, on seeing him, began again to say to the bystanders, "This man is one of them." 70 But again he denied it. Then after a little while the bystanders again said to Peter, "Certainly you are one of them; for you are a Galilean." 71 But he began to curse, and he swore an oath, "I do not know this man you are talking about." 72 At that moment the cock crowed for the second time. Then Peter remembered that Jesus had said to him, "Before the cock crows twice, you will deny me three times." And he broke down and wept.

Lk 22:55b-62

55 ... in the middle of the courtyard and sat down together, Peter sat among them. 56 Then a servant-girl came, seeing him in the firelight, stared at him and said, "This man also was with him." 57 But he denied it, saying, "Woman, I do not know him." 58 A little later someone else, on seeing him, said, "You also are one of them." But Peter said, "Man, I am not!" 59 Then about an hour later still another kept insisting, "Surely this man also was with him; for he is a Galilean." 60 But Peter said, "Man, I do not know what you are talking about!" At that moment, while he was still speaking, the cock crowed. 61 The Lord turned and looked at Peter. Then Peter remembered the word of the Lord, how he had said to him, "Before the cock crows today, you will deny me three times." 62 And he went out and wept bitterly.

Figure 7.18
Peter's Denial (NRSV translation)

Starting with the mocking of Jesus, as presented in Figures 7.15 and 7.16, we may also note the likely influence on Mark, and especially on Matthew, either directly or indirectly through the oral tradition, of a verse from one of the Suffering Servant passages in the Book of Isaiah, Isa 60:5, "I gave my back to scourges, and my cheeks to blows; and I did not turn away my face from the shame of spitting." For all the differences in detail, what is common to all the accounts is epitomised in the word προφήτευσον ("prophesy"), present in all three; but the most significant feature for the synoptic problem is the minor verbal agreement of Matthew and Luke against Mark in that they add the words τίς ἐστιν ὁ παίσας σε ("Who is it that struck you?"), absent from Mark, to indicate the content of what it is that Jesus is to prophesy. As Tuckett[49] acknowledges: "This is generally agreed to be the most perplexing of all the minor agreements." He then lends support to Streeter,[50] who argues that these words are a later interpolation into the original text of Matthew. This is an example of another strategy for explaining the minor agreements, in that sometimes there is manuscript evidence to suggest that later copyists have altered the text, often in order to harmonise the accounts of the different evangelists, so that an apparent minor agreement is in fact due to such textual corruption. The problem in the present case is that there is no manuscript evidence to support the argument. Alternatively, as always, an appeal can be made to common oral tradition to explain the agreement. However, Goulder[51] argues that Matthew has added these words in reworking the text of Mark, and that Luke then follows Matthew.

We turn now to Peter's triple denial of Jesus, with the texts in Figures 7.17 and 7.18. Again the three texts, although clearly telling the same story, parallel each other rather freely, except at the end, where with the crowing of the cock and Peter's remorse they parallel each other more closely. One of the striking differences between the gospels here is that in Mark there is reference to the cock crowing twice, whereas in Matthew and Luke the crowing occurs only once. In fact this difference goes back to an earlier point in the gospels, Mt 26:34 || Mk 14:30 || Lk 22:34, where Jesus foretells Peter's denials. In the Markan version, Jesus says: "...before the cock crows twice, you will deny me three times," but in Matthew's and Luke's parallel texts the word "twice" (δὶς) is not present. Similarly, in the present pericope, where, at Mt 26:75 || Mk 14:72 || Lk 22:61, Peter recalls Jesus' prediction, δὶς is present in Mark but not in Matthew and Luke, so that again we have a common omission. If the words at the end of Mk 14:68, καὶ ἀλέκτωρ ἐφώνησεν ("Then the cock crowed") really belong to the original text then we have another common omission by Matthew and Luke, although it is a possibility that they have

[49]Tuckett (1984), p. 136
[50]Streeter (1924), pp. 325-328
[51]Goulder (1978), pp. 226-228

been added later to the text of Mark for consistency with the references to the cock crowing twice. However, we may concur with France:[52]

> The simplest explanation, particularly for those who take seriously the tradition that Peter was himself the source of much of the material in Mark's gospel, is that Mark preserves the account in its fullest and most detailed form (as Peter himself would have remembered and repeated it), but that the vivid personal memory of the double cockcrow was omitted as an unnecessary additional detail in the other accounts.

In that case, however, it is more straightforward to envisage that either Matthew or Luke first made the decision to adopt a simpler version of the story and then the other followed suit, rather than that Matthew and Luke did so independently of each other.

As a final important minor verbal agreement of Matthew and Luke against Mark, we consider the concluding words in the account of Peter's triple denial, at Mt 26:75 || Mk 14:72 || Lk 22:62, where in Mark ἐπιβαλὼν ἔκλαιεν (difficult to translate, in the NRSV "he broke down and wept"), but where both Matthew and Luke have substituted ἐξελθὼν ἔξω ἔκλαυσεν πικρῶς ("he went out and wept bitterly"). This too is a minor agreement over which Goulder and Tuckett do battle. As far as Goulder is concerned,[53] Matthew altered the text of Mark, and Luke followed him in this. It would seem far-fetched to suppose that Matthew and Luke could have independently altered Mark's ending of the account of Peter's denials word for word in exactly the same way, even if they were influenced by similar verbal traditions; but for Tuckett,[54] following Streeter,[55] this verse may well have been missing in the original version of Luke, for which there is some very limited manuscript evidence. As remarked by Marshall,[56] the ending of Luke's account would then be similar to the parallel account in the Gospel of John, which ends at Jn 18:27 with the crowing of the cock and has no mention of Peter weeping.

In summary, we may envisage, then, that the parallel texts here have been strongly influenced by verbal traditions. However, this is not enough to account in a convincing way for all the verbal agreements of Matthew and Luke against Mark, which are most simply explained if we suppose that Matthew was also a source for Luke, or vice versa.

[52]See France (2002), p. 579. The Gospel of John also records Peter's triple denial, as predicted by Jesus, with only a single cockcrow.

[53]Goulder (1978), pp. 228-229

[54]Tuckett (1984), pp. 137-138

[55]Streeter (1924), p. 323

[56]Marshall (1978), p. 844

Table 7.9
Candidate pericopes for investigating the association between Mark and Luke in their verbal agreements with Matthew

Huck no.	Pericope description	Tables in which present
51	The Gadarene Demoniacs	6.6, 6.9, 6.10
189	The Rich Young Man	6.9, 6.10 (also 5.7)
193	The Healing of Bartimaeus	6.6, 6.10

7.8 Comments on further pericopes

In Tables 6.6, 6.9 and 6.10 of Chapter 6 we identified passages that might be of special interest if we wished to investigate the association between Mark and Luke in their verbal agreements with Matthew, or more specifically, under the assumption of Matthean priority, the association between Mark and Luke in their use of Matthew. Although this issue is not of direct concern for examination of the two-source hypothesis, it may be of interest to see if these passages serve to illuminate any other aspects of synoptic relationships. On the assumption of Markan priority, these turn out to be passages such that Luke follows Mark more closely than does Matthew. They stand out as unusual, apart from anything else, because overall Matthew follows Mark more closely than does Luke. To have a small number of sample passages to look at, we list in Table 7.9 those pericopes that appear in at least two of the above tables. In addition we note that the Huck pericope number 189 also appears in Table 5.7, suggesting a strong positive association between Luke's and Matthew's use of Mark in this pericope, which provides an additional reason for examining it in detail.

Assuming the priority of Mark, the first of the passages in Table 7.9, the pericope of the Gadarene demoniacs in Matthew, is a much shortened version of the corresponding pericope in Mark, but Luke has a longer version than Matthew, following Mark more closely. The parallels are Mt 8:28-34 || Mk 5:1-20 || Lk 8:26-39. Because in Matthew the pericope occurs earlier in the sequence of pericopes than it does in Mark and Luke, it appears twice in the Huck synopsis, as pericope number 51 ("The Gadarene Demoniacs"), corresponding to its position in Matthew, and as pericope number 106 ("The Gerasene Demoniac"), corresponding to its position in Mark and Luke. We shall not examine in detail the text of what in Mark and Luke is a longish pericope, but comment briefly on some differences between Matthew's version on the one hand and Mark's and Luke's on the other. It is noteworthy that, although there are significant variations in the manuscript tradition, the Greek of NA^{25} and of the later critical editions has Matthew placing the exorcism in "the country of the Gadarenes", whereas Mark and Luke have it in "the country of the Gerasenes". Another immediate difference is that Matthew has two men possessed by demons, whereas Mark and Luke have only one. This may suggest that Matthew is using or is influenced by an oral tradition of the same exorcism that differs substantially from the tradition that underlies the texts of Mark and Luke.

But the pericope is picked up in our analyses in Chapter 6 because there is a strong association between Mark and Luke in their verbal agreements and non-agreements with Matthew. Specifically, there are many verbal agreements of Mark and Luke against Matthew. Under the hypothesis of Markan priority, this has little impact on the issue of whether Matthew and Luke were independent in their use of Mark. However, it is an important pericope to study for any hypothesis that involves the priority of Matthew — it strongly indicates that Luke also had Mark as a source or that Mark also had Luke as a source, or perhaps that Mark and Luke were independently influenced by similar oral traditions or some other source material. According to McNicol at al. and Peabody et al.,[57] who advocate the Griesbach/two-gospel hypothesis, it is just a matter of Luke altering and elaborating upon the text of Matthew and then Mark following Luke.

In the following sections we shall examine in more detail the other two pieces of text indicated in Table 7.9, under the assumption of Markan priority.

7.9 The Rich Young Man

The section of text Mt 19:16-21 was picked out in Table 6.9, the section Mt 19:13-21 in Table 6.10 and the section Mk 10:13-28 in Table 5.7. They fall within the Huck pericope number 188, Mt 19:13-15 || Mk 10:13-16 || Lk 18:15-17, "Jesus Blesses the Children", and the Huck pericope number 189, Mt 19:16-30 || Mk 10:17-31 || Lk 18:18-30, "The Rich Young Man". We have selected for more detailed discussion the intersection of the three suggested sections of text, Mt 19:16-21 || Mk 10:17-21 || Lk 18:18-22, which forms the first part of the pericope of the Rich Young Man. In Figures 7.19 and 7.20 we give the parallel texts in Greek and English, respectively.

In Table 7.10 we present the counts of verbal agreements of Matthew and Luke with Mark, aggregated over the 97 words in the text of Mk 10:17-21, and in brackets the expected frequencies, given the row and column totals, under the hypothesis that Matthew and Luke are statistically independent in their verbal agreements with Mark. As in all the previous examples, the observed frequencies along the diagonal of the table are greater than the ex-

[57]McNicol et al. (1996), p. 129; Peabody et al. (2002), pp. 140-144

Mt 19:16-21

16 Καὶ ἰδοὺ εἷς προσελθὼν αὐτῷ εἶπεν· διδάσκαλε, τί ἀγαθὸν ποιήσω ἵνα σχῶ ζωὴν αἰώνιον; 17 ὁ δὲ εἶπεν αὐτῷ· τί με ἐρωτᾷς περὶ τοῦ ἀγαθοῦ; εἷς ἐστιν ὁ ἀγαθός· εἰ δὲ θέλεις εἰς τὴν ζωὴν εἰσελθεῖν, τήρει τὰς ἐντολάς. 18 Λέγει αὐτῷ· ποίας; ὁ δὲ Ἰησοῦς ἔφη· τὸ οὐ φονεύσεις, οὐ μοιχεύσεις, οὐ κλέψεις, οὐ ψευδομαρτυρήσεις, 19 τίμα τὸν πατέρα καὶ τὴν μητέρα, καὶ ἀγαπήσεις τὸν πλησίον σου ὡς σεαυτόν. 20 λέγει αὐτῷ ὁ νεανίσκος· ταῦτα πάντα ἐφύλαξα· τί ἔτι ὑστερῶ; 21 ἔφη αὐτῷ ὁ Ἰησοῦς· εἰ θέλεις τέλειος εἶναι, ὕπαγε πώλησόν σου τὰ ὑπάρχοντα καὶ δὸς πτωχοῖς, καὶ ἕξεις θησαυρὸν ἐν οὐρανοῖς, καὶ δεῦρο ἀκολούθει μοι.

Mk 10:17-21

17 Καὶ ἐκπορευομένου αὐτοῦ εἰς ὁδὸν προσδραμὼν εἷς καὶ γονυπετήσας αὐτὸν ἐπηρώτα αὐτόν· διδάσκαλε ἀγαθέ, τί ποιήσω ἵνα ζωὴν αἰώνιον κληρονομήσω; 18 ὁ δὲ Ἰησοῦς εἶπεν αὐτῷ· τί με λέγεις ἀγαθόν; οὐδεὶς ἀγαθὸς εἰ μὴ εἷς ὁ θεός. 19 τὰς ἐντολὰς οἶδας· μὴ φονεύσῃς, μὴ μοιχεύσῃς, μὴ κλέψῃς, μὴ ψευδομαρτυρήσῃς, μὴ ἀποστερήσῃς, τίμα τὸν πατέρα σου καὶ τὴν μητέρα. 20 ὁ δὲ ἔφη αὐτῷ· διδάσκαλε, ταῦτα πάντα ἐφυλαξάμην ἐκ νεότητός μου. 21 ὁ δὲ Ἰησοῦς ἐμβλέψας αὐτῷ ἠγάπησεν αὐτὸν καὶ εἶπεν αὐτῷ· ἕν σε ὑστερεῖ· ὕπαγε, ὅσα ἔχεις πώλησον καὶ δὸς τοῖς πτωχοῖς, καὶ ἕξεις θησαυρὸν ἐν οὐρανῷ, καὶ δεῦρο ἀκολούθει μοι.

Lk 18:18-22

18 Καὶ ἐπηρώτησέν τις αὐτὸν ἄρχων λέγων· διδάσκαλε ἀγαθέ, τί ποιήσας ζωὴν αἰώνιον κληρονομήσω; 19 εἶπεν δὲ αὐτῷ ὁ Ἰησοῦς· τί με λέγεις ἀγαθόν; οὐδεὶς ἀγαθὸς εἰ μὴ εἷς ὁ θεός. 20 τὰς ἐντολὰς οἶδας· μὴ μοιχεύσῃς, μὴ φονεύσῃς, μὴ κλέψῃς, μὴ ψευδομαρτυρήσῃς, τίμα τὸν πατέρα σου καὶ τὴν μητέρα. 21 ὁ δὲ εἶπεν· ταῦτα πάντα ἐφύλαξα ἐκ νεότητος. 22 ἀκούσας δὲ ὁ Ἰησοῦς εἶπεν αὐτῷ· ἔτι ἕν σοι λείπει· πάντα ὅσα ἔχεις πώλησον καὶ διάδος πτωχοῖς, καὶ ἕξεις θησαυρὸν ἐν τοῖς οὐρανοῖς, καὶ δεῦρο ἀκολούθει μοι.

Figure 7.19
The Rich Young Man (NA[25] Greek)

Table 7.10
Counts of verbal agreements with Mk 10:17-21

		Luke		
		0	1	total
Matthew	0	24 (16.08)	28 (35.92)	52
	1	6 (13.92)	39 (31.08)	45
	total	30	67	97

Mt 19:16-21	Mk 10:17-21	Lk 18:18-22
16 Then someone came to him and said, "Teacher, what good deed must I do to have eternal life?" 17 And he said to him, "Why do you ask me about what is good? There is only one who is good. If you wish to enter into life, keep the commandments." 18 He said to him, "Which ones?" And Jesus said, "You shall not murder; You shall not commit adultery; You shall not steal; You shall not bear false witness; 19 Honour your father and mother; also, You shall love your neighbour as yourself." 20 The young man said to him, "I have kept all these; what do I still lack?" 21 Jesus said to him, "If you wish to be perfect, go, sell your possessions, and give the money to the poor, and you will have treasure in heaven; then come, follow me."	17 As he was setting out on a journey, a man ran up and knelt before him, and asked him, "Good Teacher, what must I do to inherit eternal life?" 18 Jesus said to him, "Why do you call me good? No one is good but God alone. 19 You know the commandments: 'You shall not murder; You shall not commit adultery; You shall not steal; You shall not bear false witness; You shall not defraud; Honour your father and mother.' " 20 He said to him, "Teacher, I have kept all these since my youth." 21 Jesus, looking at him, loved him and said, "You lack one thing; go, sell what you own, and give the money to the poor, and you will have treasure in heaven; then come, follow me."	18 A certain ruler asked him, "Good Teacher, what must I do to inherit eternal life?" 19 Jesus said to him, "Why do you call me good? No one is good but God alone. 20 You know the commandments: 'You shall not commit adultery; You shall not murder; You shall not steal; You shall not bear false witness; Honour your father and mother.' " 21 He replied, "I have kept all these since my youth." 22 When Jesus heard this, he said to him, "There is still one thing lacking. Sell all that you own and distribute the money to the poor, and you will have treasure in heaven; then come, follow me."

Figure 7.20
The Rich Young Man (NRSV translation)

pected frequencies. Yet again there appears to be a positive association between Matthew and Luke in their verbal agreements with Mark. We may also observe in Figure 7.19 the 3 boxed words that are minor verbal agreements of Matthew and Luke against Mark. However, the feature of Table 7.10 that makes it markedly different from the similar tables for the sections of text examined earlier is in the large numbers of agreements between Luke and Mark, 67 out of 97 words, especially as compared with the agreements between Matthew and Mark, 45 out of 97 words. Furthermore, there are 28 verbal agreements of Mark and Luke against Matthew, in terms of Figure 7.19, 28 words with a wavy underline, a much larger number than in the previous examples. This feature to a large extent explains why this section of text was picked out for investigation in Tables 6.9 and 6.10.

For Davies and Allison,[58] Matthew's version of this material is based on Mark's, but Matthew has undertaken substantial rewriting. For example, in Mk 10:17-18 || Lk 18:18-19 Jesus is addressed as "Good Teacher" and replies "Why do you call me good? No one is good but God alone". This has been seen as a source of embarrassment by commentators, as it might be taken to suggest that Jesus was not absolutely good, that he was tainted by sin. Assuming Markan priority, it seems plausible, as suggested by Davies and Allison, that Matthew too would have seen this as a difficulty and therefore rewritten this material. At Mt 19:16-17 he has moved the word "good" from the address "Good Teacher" and has instead "...what good deed must I do ..." with Jesus' reply "Why do you ask me about what is good? ..." However, Luke here follows Mark almost word for word.

At Mt 19:18-19 || Mk 10:19 || Lk 18:20 there is a quotation from the latter part of the ten commandments, as found both at Exodus 20:12-17 and at Deuteronomy 5:16-21, where the Septuagint reads:

> 16 τίμα τὸν πατέρα σου καὶ τὴν μητέρα σου ...17 οὐ μοιχεύσεις. 18 οὐ φονεύσεις. 19 οὐ κλέψεις. 20 οὐ ψευδομαρτυρήσεις κατὰ τοῦ πλησίον σου μαρτυρίαν ψευδῆ. 21 οὐκ ἐπιθυμήσεις ...
>
> 16 Honour your father and your mother ...17 You shall not commit adultery. 18 You shall not murder. 19 You shall not steal. 20 You shall not testify falsely against your neighbour with a false testimony. 21 You shall not covet ...

As we saw in Section 7.5, there may be considerable variability within scriptural quotations, and so it is here. Assuming Markan priority, Mark has moved the commandment to "honour your father and mother" to the end of the quotation, and in this he is followed by both Matthew and Luke. Mark has a simpler version of the commandment "You shall not testify falsely ...", in which also he is followed by both Matthew and Luke. Furthermore, Mark has not included the commandment "You shall not covet ...". Intriguingly,

[58] Davies and Allison (1997), Volume III, pp. 39-40

he has instead a commandment μὴ ἀποστερήσῃς ("You shall not defraud"), which does not appear in the Old Testament scripture, but here there is a minor agreement between Matthew and Luke against Mark in that they both omit this commandment. It does seem plausible that Matthew and Luke could have made the omission independently of each other if this commandment did not appear in the scriptures as they knew them. Matthew alone, however, adds the commandment "You shall love your neighbour as yourself" from Leviticus 19:18, which as we saw in Section 7.5 appears also at Mt 22:39 || Mk 12:31 || Lk 10:27.

In the list of prohibitions, Mark and Matthew have murder ahead of adultery, as in the Hebrew bible, whereas Luke has adultery ahead of murder, as in the Septuagint, so that there is an agreement of Mark and Matthew against Luke. However, there is also considerable textual support for the order with adultery ahead of murder in Mark, which Gundry[59] suggests is more likely to be the original order in Mark, the change in order having occurred through assimilation to the parallel text in Matthew. In that case, the agreement would be of Mark and Luke against Matthew, as it tends to be more generally in this pericope. In the list of prohibitions as a whole, Mark and Luke use the particle μὴ with the verb in the subjunctive, whereas Matthew differs in that, in conformity with the Septuagint text, he uses the particle οὐ with the verb in the future indicative. This is a difference that stands out in the Greek but is not apparent in the English translation.

Besides the omission of the commandment "You shall not defraud" in Mk 10:19, there are a few other minor agreements of Matthew and Luke against Mark, notably the common omission of ἐμβλέψας αὐτῷ ἠγάπησεν αὐτὸν ("looking at him, loved him") in Mk 10:21a. It may be recalled from Section 7.2 that in our first example of the Healing of a Leper there were two instances of common omission of words that described the emotions of Jesus. Here too, though perhaps for different reasons, it may be argued that both Matthew and Luke, independently of each other, did not wish to attribute the emotion of love to Jesus.[60] However, the strength of any such argument is open to question, and it seems more likely that Luke was influenced by Matthew or vice versa. There are also the three minor verbal agreements, the three words shown as boxed text in Figure 7.19: both Matthew and Luke have changed the middle voice ἐφυλαξάμην ("I have kept") in Mk 10:20 to the more natural active voice ἐφύλαξα; both Matthew and Luke have inserted ἔτι ("still"); and in Mk 10:21b, where Matthew and Luke are otherwise following Mark almost word for word, they have both changed the singular οὐρανῷ ("heaven") to the plural οὐρανοῖς. How significant all these minor agreements are is debatable, although on the whole it seems unlikely that Matthew and Luke should have made all these common alterations to the text of Mark independently of each

[59]Gundry (1994), p. 386

[60]Davies and Allison (1997), Volume III, p. 46, n. 54, put forward the possibility that Matthew and Luke might have wished to remove any suggestion of homosexuality.

other. For Gundry,[61] "With other agreements between Matthew and Luke against Mark, Luke's omission of Jesus' love suggests Matthean influence." On the other hand, Goulder, who otherwise, in his advocacy of the Farrer hypothesis, argues for Luke's dependency on Matthew, is cautious here,[62] noting that "In the main Luke follows the Marcan text faithfully" and that "Luke has followed Mark without reference to Matthew ... ", although he does also see some evidence for the influence of Matthew on Luke.

In summary, then, on the assumption of Markan priority, a very striking feature of the present passage is that Luke certainly followed Mark much more closely than did Matthew. On the assumption that Luke's was the last of the synoptic gospels to be written, there is some evidence, though not as strong as in the earlier examples of Sections 7.2 – 7.7, that, in his use of Mark, Luke was influenced by Matthew. This tallies with the fact that this passage appeared in Tables 6.9 and 6.10 with regard to a positive association between Mark and Luke in their verbal agreements and non-agreements with Matthew, but only in Table 5.7 with regard to a positive association between Luke and Matthew in their verbal agreements and non-agreements with Mark.

Naturally, for McNicol at al. and Peabody et al.,[63] with their advocacy of the Griesbach hypothesis, Luke has revised the text of Matthew, and Mark has followed Luke's revision.

7.10 The Healing of Bartimaeus

In Tables 6.6 and 6.10 the whole of the Huck pericope number 193, Mt 20:29-34 || Mk 10:46-52 || Lk 18:35-43, "The Healing of Bartimaeus", was picked out for further investigation. In Figures 7.21 and 7.22 we give the parallel texts in Greek and English, respectively.

A complicating feature here is that there is what is generally regarded as an earlier Matthean version, a doublet, of this pericope at Mt 9:27-31, Huck pericope number 56, "Two Blind Men Healed" – not present in Mark or Luke. Although it is certainly worth being aware of the first version for any input it may provide in discussing Mt 20:29-34, it is this second version that is taken to be parallel to the corresponding accounts in Mark and Luke and used in ascertaining and counting the verbal agreements. Furthermore, it is worth keeping in mind the pericope at Mk 8:22-26, Huck pericope number 121, "The Blind Man of Bethsaida", which is one of the few single-tradition Markan pericopes, i.e., material unique to Mark,[64] but which, as we shall see,

[61] Gundry (1994), p. 388
[62] Goulder (1989), p. 671
[63] McNicol et al. (1996), pp. 237-238; Peabody et al. (2002), pp. 220-222
[64] There is, however, an interesting parallel to Mk 8:22-26 in the Gospel of John, at Jn 9:1-7.

Table 7.11
Counts of verbal agreements with Mk 10:46-52

		Luke		
		0	1	total
Matthew	0	59 (50.44)	35 (43.56)	94
	1	7 (15.56)	22 (13.44)	29
	total	66	57	123

may have influenced Matthew, even though he did not reproduce it in his gospel.

In Table 7.11 we present the counts of verbal agreements of Matthew and Luke with Mark, aggregated over the 123 words in the text of Mk 10:46-52, and in brackets the expected frequencies, given the row and column totals, under the hypothesis that Matthew and Luke are statistically independent in their verbal agreements with Mark. The observed frequencies along the diagonal of the table are, as in all our previous examples, greater than the expected frequencies, so that again there appears to be some positive association between Matthew and Luke in their verbal agreements with Mark. In Figure 7.21 we see that there is a single boxed word κύριε ("Lord"), a minor verbal agreement of Matthew and Luke against Mark. In comparison with Table 7.10 in the previous section, there are in Table 7.11 proportionally fewer verbal agreements of Matthew and Luke with Mark. However, just as in Table 7.10, we see in Table 7.11 that there is a large number of verbal agreements of Mark and Luke against Matthew, 35 in total, in terms of Figure 7.21, 35 words with a wavy underline. Under the assumption of Markan priority, neither Matthew nor Luke is following Mark very closely, but Luke is following Mark much more closely than is Matthew.

An immediately obvious difference between Matthew on the one hand and Mark and Luke on the other is that Matthew has two blind men being healed, whereas Mark and Luke have only one.[65] Matthew also differs in a significant way from Mark and Luke at Mt 20:34a || Mk 10:52a || Lk 18:42. Firstly, Matthew has inserted the word σπλαγχνισθεὶς ("moved with compassion"), which is especially noteworthy because, at Mt 8:3, in the pericope of the Healing of a Leper, Matthew, along with Luke, omitted rather than inserted

[65]This situation is reminiscent of the pericope of the Gadarene Demoniacs looked at briefly in Section 7.8, where Matthew has two men possessed by demons, but Mark and Luke in their parallels have only one.

Mt 20:29-34	Mk 10:46-52	Lk 18:35-43
29 Καὶ ἐκπορευομένων αὐτῶν ἀπὸ Ἰεριχὼ ἠκολούθησεν αὐτῷ ὄχλος πολύς. 30 καὶ ἰδοὺ δύο τυφλοὶ καθήμενοι παρὰ τὴν ὁδόν, ἀκούσαντες ὅτι Ἰησοῦς παράγει, ἔκραξαν λέγοντες· κύριε, ἐλέησον ἡμᾶς, υἱὸς Δαυίδ. 31 ὁ δὲ ὄχλος ἐπετίμησεν αὐτοῖς ἵνα σιωπήσωσιν· οἱ δὲ μεῖζον ἔκραξαν λέγοντες· κύριε, ἐλέησον ἡμᾶς, υἱὸς Δαυίδ. 32 καὶ στὰς ὁ Ἰησοῦς ἐφώνησεν αὐτοὺς καὶ εἶπεν· τί θέλετε ποιήσω ὑμῖν; 33 λέγουσιν αὐτῷ· κύριε, ἵνα ἀνοιγῶσιν οἱ ὀφθαλμοὶ ἡμῶν. 34 σπλαγχνισθεὶς δὲ ὁ Ἰησοῦς ἥψατο τῶν ὀμμάτων αὐτῶν, καὶ εὐθέως ἀνέβλεψαν καὶ ἠκολούθησαν αὐτῷ.	46 Καὶ ἔρχονται εἰς Ἰεριχώ. Καὶ ἐκπορευομένου αὐτοῦ ἀπὸ Ἰεριχὼ καὶ τῶν μαθητῶν αὐτοῦ καὶ ὄχλου ἱκανοῦ ὁ υἱὸς Τιμαίου Βαρτιμαῖος, τυφλὸς προσαίτης, ἐκάθητο παρὰ τὴν ὁδόν. 47 καὶ ἀκούσας ὅτι Ἰησοῦς ὁ Ναζαρηνός ἐστιν ἤρξατο κράζειν καὶ λέγειν· υἱὲ Δαυὶδ Ἰησοῦ, ἐλέησόν με. 48 καὶ ἐπετίμων αὐτῷ πολλοὶ ἵνα σιωπήσῃ· ὁ δὲ πολλῷ μᾶλλον ἔκραζεν· υἱὲ Δαυίδ, ἐλέησόν με. 49 καὶ στὰς ὁ Ἰησοῦς εἶπεν· φωνήσατε αὐτόν. καὶ φωνοῦσιν τὸν τυφλὸν λέγοντες αὐτῷ· θάρσει, ἔγειρε, φωνεῖ σε. 50 ὁ δὲ ἀποβαλὼν τὸ ἱμάτιον αὐτοῦ ἀναπηδήσας ἦλθεν πρὸς τὸν Ἰησοῦν. 51 καὶ ἀποκριθεὶς αὐτῷ ὁ Ἰησοῦς εἶπεν· τί σοι θέλεις ποιήσω; ὁ δὲ τυφλὸς εἶπεν αὐτῷ· ῥαββουνί, ἵνα ἀναβλέψω. 52 καὶ ὁ Ἰησοῦς εἶπεν αὐτῷ· ὕπαγε, ἡ πίστις σου σέσωκέν σε. καὶ εὐθὺς ἀνέβλεψεν καὶ ἠκολούθει αὐτῷ ἐν τῇ ὁδῷ.	35 Ἐγένετο δὲ ἐν τῷ ἐγγίζειν αὐτὸν εἰς Ἰεριχὼ τυφλός τις ἐκάθητο παρὰ τὴν ὁδὸν ἐπαιτῶν. 36 ἀκούσας δὲ ὄχλου διαπορευομένου ἐπυνθάνετο τί εἴη τοῦτο. 37 ἀπήγγειλαν δὲ αὐτῷ ὅτι Ἰησοῦς ὁ Ναζωραῖος παρέρχεται. 38 καὶ ἐβόησεν λέγων· Ἰησοῦ υἱὲ Δαυίδ, ἐλέησόν με. 39 καὶ οἱ προάγοντες ἐπετίμων αὐτῷ ἵνα σιγήσῃ, αὐτὸς δὲ πολλῷ μᾶλλον ἔκραζεν· υἱὲ Δαυίδ, ἐλέησόν με. 40 σταθεὶς δὲ ὁ Ἰησοῦς ἐκέλευσεν αὐτὸν ἀχθῆναι πρὸς αὐτόν. ἐγγίσαντος δὲ αὐτοῦ ἐπηρώτησεν αὐτόν· 41 τί σοι θέλεις ποιήσω; ὁ δὲ εἶπεν· κύριε, ἵνα ἀναβλέψω. 42 καὶ ὁ Ἰησοῦς εἶπεν αὐτῷ· ἀνάβλεψον· ἡ πίστις σου σέσωκέν σε. 43 καὶ παραχρῆμα ἀνέβλεψεν καὶ ἠκολούθει αὐτῷ δοξάζων τὸν θεόν. καὶ πᾶς ὁ λαὸς ἰδὼν ἔδωκεν αἶνον τῷ θεῷ.

Figure 7.21
The Healing of Bartimaeus (NA^{25} Greek)

Mt 20:29-34	Mk 10:46-52	Lk 18:35-43
29 As they were leaving Jericho, a large crowd followed him. 30 There were two blind men sitting by the roadside. When they heard that Jesus was passing by, they shouted, "Lord, have mercy on us, Son of David!" 31 The crowd sternly ordered them to be quiet; but they shouted even more loudly, "Have mercy on us, Lord, Son of David!" 32 Jesus stood still and called them, saying, "What do you want me to do for you?" 33 They said to him, "Lord, let our eyes be opened." 34 Moved with compassion, Jesus touched their eyes. Immediately they regained their sight and followed him.	46 They came to Jericho. As he and his disciples and a large crowd were leaving Jericho, Bartimaeus son of Timaeus, a blind beggar, was sitting by the roadside. 47 When he heard that it was Jesus of Nazareth, he began to shout out and say, "Jesus, Son of David, have mercy on me!" 48 Many sternly ordered him to be quiet, but he cried out even more loudly, "Son of David, have mercy on me!" 49 Jesus stood still and said, "Call him here." And they called the blind man, saying to him, "Take heart; get up, he is calling you." 50 So throwing off his cloak, he sprang up and came to Jesus. 51 Then Jesus said to him, "What do you want me to do for you?" The blind man said to him, "My teacher, let me see again." 52 Jesus said to him, "Go; your faith has made you well." Immediately he regained his sight and followed him on the way.	35 As he approached Jericho, a blind man was sitting by the roadside begging. 36 When he heard a crowd going by, he asked what was happening. 37 They told him, "Jesus of Nazareth is passing by." 38 Then he shouted, "Jesus, Son of David, have mercy on me!" 39 Those who were in front sternly ordered him to be quiet; but he shouted even more loudly, "Son of David, have mercy on me!" 40 Jesus stood still and ordered the man to be brought to him; and when he came near, he asked him, 41 "What do you want me to do for you?" He said, "Lord, let me see again." 42 Jesus said to him, "Receive your sight; your faith has saved you." 43 Immediately he regained his sight and followed him, glorifying God; and all the people, when they saw it, praised God.

Figure 7.22
The Healing of Bartimaeus (NRSV translation)

σπλαγχνισθεὶς.[66] Furthermore, Matthew omits the words ἡ πίστις σου σέσωκέν σε ("your faith has made you well/saved you"), present in Mark and Luke. Instead, Matthew describes Jesus touching the eyes of the blind men, thus emphasising the compassion and action of Jesus, as against the faith of the blind man that is brought out in Mark and Luke. For both Gundry and Davies and Allison[67] this suggests the influence on Matthew of the Markan pericope Mk 8:22-26 with its strong emphasis on the physical aspects of the healing of the blind man. This suggestion is reinforced by the switch from the use of the word ὀφθαλμός for "eye" at Mt 20:33 to the synonymous, but much rarer, ὄμμα at Mt 20:34. The only other place in the *New Testament* where the word ὄμμα is used is at Mk 8:23, with a switch to ὀφθαλμός at Mk 8:25. Matthew has, however, already emphasised the faith of the blind men in the doublet at Mt 9:27-31.

There is also a notable agreement of Matthew and Mark against Luke. All three evangelists agree that the healing took place at Jericho, but, according to Matthew and Mark, it took place when Jesus and the crowd were leaving Jericho, whereas, according to Luke, it took place when Jesus and the crowd were entering Jericho (Mt 20:29 || Mk 10:46 || Lk 18:35). This divergence from Matthew and Mark, presumably due to Lukan redaction, allows Luke to locate his following single-tradition story of the tax collector Zacchaeus in Jericho.

However, some of the main elements of Mark's account are retained more or less unchanged, with minor variations in wording, by both Matthew and Luke. The healing took place in Jericho, where the blind man/men were sitting παρὰ τὴν ὁδὸν ("by the roadside"). When he/they heard that Jesus was passing by, he/they cried out once and then again a second time, even more loudly, "Son of David, have mercy on me/us". Jesus stopped and asked him/them what they wanted him to do for him/them. Jesus healed him/them so that immediately he/they regained their sight and followed him.

There are some minor agreements of Matthew and Luke against Mark, though not as many as in our earlier examples in Sections 7.2 – 7.7. Notably, there is the common omission of most of Mk 10:49-50, where in Mark the calling of the blind man by the crowd, his jumping up and his coming to Jesus is described in more colourful detail than is found in Matthew or Luke; and there is also the common omission of the phrase ἐν τῇ ὁδῷ ("on the way") at the end of Mk 10:52. Both Matthew and Luke omit the name of the beggar, Bartimaeus son of Timaeus, that is given in Mk 10:46, and both Matthew and Luke replace the blind man's address of Jesus at Mk 10:51 as ῥαββουνί ("my teacher") by κύριε ("Lord"). These minor agreements are for Davies and Allison "insignificant".[68] Certainly it seems plausible that Matthew and Luke could have independently replaced the Aramaic word ῥαββουνί, which occurs only twice in the *New Testament*, here and at John 20:16, with the commonly used Greek title κύριε for Jesus. However, it is more difficult to envisage that

[66]See Section 7.2
[67]Gundry (1994), p. 406; Davies and Allison (1997), Volume III, pp. 104, 108
[68]Davies and Allison (1997), Volume III, p. 105

Matthew and Luke would have, independently of each other, made all the common omissions noted above.

We may also observe that Matthew has inserted κύριε, a characteristic of Matthew's style,[69] to give the liturgical phrase κύριε ἐλέησον ("Lord have mercy")[70] in two other places in this pericope, where the blind men are shouting out to Jesus; but Luke has not, so that these are negative agreements of Mark and Luke against Matthew.

It should be recalled that this pericope was not picked out for investigation because it might provide another example, under the assumption of Markan priority, of evidence of dependence between Matthew and Luke in their use of Mark, and so it should not be surprising that, even if some such evidence may be detected, it is not as clear-cut as we have seen elsewhere. This pericope was picked out, instead, because of the strong statistical dependence between Mark and Luke in their verbal agreements and non-agreements with Matthew. As we have seen, there are notable agreements of each pair of gospels against the third, but the most fundamental are the agreements of Mark and Luke against Matthew. The fact that there are common core elements in the three versions of this healing story, but that Matthew also diverges substantially from Mark and Luke, may suggest that Matthew is being influenced by a variant oral tradition. In any case, this pericope provides a salutary reminder of some of the complexities of the relationships between the synoptic gospels, of which the synoptic models that we have been considering, especially in the earlier chapters, are able to provide only a partial account.

McNicol at al. and Peabody et al.,[71] with their assumption of Matthean priority, devote considerable attention to this pericope. According to McNicol at al., Luke has modified Mt 20:29-34 by conflating it with the doublet at Mt 9:27-31; and, according to Peabody et al., Mark uses Mt 20:29-34 and Lk 18:35-43, alternating between them in his agreements and adding his own supplementary material.

7.11 Conclusions

It does appear that our earlier statistical analysis has been fruitful in leading to a systematic way of identifying passages for detailed investigation. It has turned out that we have indeed been able to focus on a variety of passages, the ones discussed in Sections 7.2 – 7.7, that are of particular interest for probing the two-document hypothesis, specifically the assumption that neither Luke nor Matthew had the other as a source.

For defenders of the two-document hypothesis, we have seen that there

[69]See the comments in Section 7.2 on κύριε as a characteristic of Matthew's style.
[70]See Davies and Allison (1997), Volume III, p. 107
[71]McNicol et al. (1996), pp. 240-241; Peabody et al. (2002), pp. 230-234

is a variety of ways of explaining away the statistical dependence between Matthew and Luke in their use of Mark.

1. There may be specific context-dependent reasons why Matthew and Luke would in some instances concur in keeping the wording of Mark's text unchanged but in others concur in altering the text or omitting it altogether. Matthew and Luke may independently be employing similar editorial criteria, for example, in making stylistic improvements or removing what they perceive to be difficult or redundant pieces of text from Mark.

2. Where there are agreements of Matthew and Luke against Mark, it may be because both Matthew and Luke are being influenced by the same or similar oral traditions. But such an appeal to oral tradition tends to be rather vague, in that it is not spelt out just how the transmission of the tradition and its impact on the gospel writers is envisaged. If, for instance, it is supposed that the gospel writers were working in situations where there was considerable variability in the oral tradition, a more convincing case would be made if some suggested explanation was provided of how it might have come about that Matthew and Luke were influenced in a similar way, but one that differed from Mark. As we noted in Section 1.3, oral traditions must have played an important role in the emergence of the gospel texts, but discussion of the precise way in which this occurred is highly speculative.

3. Another explanation that applies in some instances is that there may have been alteration of the text by later copyists, especially in order to harmonize the different gospel accounts by assimilating one gospel text to a parallel one in another gospel. But the cases where there is strong evidence for such or other types of textual corruption are relatively few.

When we consider the totality of the evidence of statistical dependence between Matthew's and Luke's use of Mark and the variety of the ways, often requiring considerable ingenuity, that are used by supporters of the two-source hypothesis to explain the dependence in each individual case, it is tempting, to say the least, to opt for the far simpler explanation that Luke may simply be using Matthew as a source, or Matthew may be using Luke as a source. If we take that option, it raises the question of whether we can dispense with the hypothesis of a written Q source altogether. The answer to that question will be positive if, for example, we adopt the Farrer hypothesis (Figure 1.3), but negative if we adopt the three-source hypothesis (Figure 3.1). Our analysis here does not help us in any immediate way to choose between these two alternatives.

We have seen that there are aspects of the evidence pertaining to the synoptic problem that are not covered by our statistics of verbal agreements, for example, issues of differences in order, both at the micro level of word order and at the macro level of the order of the pericopes. Another issue is the way in which each of the gospel authors handles scriptural quotations, specifically with reference to the Septuagint. For the future, there is much potential for exploring the construction and analysis of data sets that go beyond the mere recording of verbal agreements and non-agreements as carried out in this study.

In the final three examples of parallel texts, studied in Sections 7.8 – 7.10, there come into view some other aspects of the synoptic problem that lie beyond our primary focus here, which is on the issue of the apparent dependence between Matthew and Luke in their use of Mark. Following on from the material in Sections 7.2 – 7.7, they provide further illustrations of the complexities of synoptic relationships and pointers to issues that might be addressed in the construction of more sophisticated statistical models. For example, it would be worth considering whether a way might be found of explicitly taking into account the role of other pericopes, especially those that may be regarded as doublets, in influencing how a particular pericope in an earlier gospel is used in a later gospel.

8

Final conclusions

8.1 Summary

This monograph has presented the results of a project carried out over the last ten years or so to investigate statistical approaches to the synoptic problem based on the data of verbal agreements. We summarise below its main themes and conclusions.

In Chapter 2 we investigated versions of Honoré's triple-link model for synoptic relationships, using data that consisted of overall counts of verbal agreements. After correcting Honoré's own analysis, we found that the triple-link model, whether in its original or its modified form, on its own terms appeared to fit the data reasonably well, although no formal statistical tests of goodness of fit could be carried out. None of the cases of the model, in whatever order the gospels were assumed to have been produced, could clearly be ruled out, but the analysis found that the cases with Luke as the last of the synoptic gospels to be written tended to give the best fit and the cases with Mark as the last gave the least good fit, which could, with appropriate qualifications, be regarded as evidence against the Griesbach hypothesis. Overall we could conclude that we had found a class of models that appeared to fit the data well, while not requiring the use of a Q source.

We noted that the triple-link model did not include within its scope the commonly adopted two-source hypothesis, with Mark and Q as sources, and in Chapters 3 and 5, by investigation of the use of Mark by Matthew and Luke, we turned to a particular aspect of this hypothesis. Through the statistical analysis of binary time series, constructed using the text of Mark as the base text and recording the word-by-word verbal agreements of Matthew and Luke with Mark, we could focus especially on the question of whether Matthew and Luke proceeded independently in their use of Mark, as is the case according to the two-source hypothesis. From the analysis of the logistic regression models in Chapter 3 it was found that there was very strong, statistically significant evidence that Matthew and Luke were not statistically independent in their verbal agreements with Mark. An important by-product of the analysis in Chapter 3 and of the fitting of hidden Markov models in Chapter 5 was the identification of a number of passages where, on the face of it, the dependence between Matthew's and Luke's use of Mark appeared to be particularly strong.

Some of the passages so marked out were subjected to detailed textual analysis in Chapter 7.

At this stage the over-arching structure of the project became explicit. Starting with the Greek texts of the synoptic gospels, statistical data sets had been constructed and analysed. Through the statistical analysis, passages had been identified that appeared likely to be of special interest for study of the synoptic problem, specifically with regard to the statistical dependence between Matthew and Luke in their verbal agreements with Mark. Finally, in Chapter 7, we were able to return from what to some may seem the barren abstractions of the statistical analyses, in particular those of the binary time series, to the far richer vein of the gospel texts themselves. The strongly inter-disciplinary nature of this work also became apparent. Whereas in earlier chapters extensive statistical analyses had been carried out, some of them, such as those involving hidden Markov models, using quite advanced techniques and requiring substantial computational work, in Chapter 7 the analyses of the texts were of a type that normally lie within the domain of the New Testament scholar.

We found that the sections of text suggested for detailed examination by the statistical analyses of the earlier chapters did indeed turn out to be fruitful for discussion of Matthew's and Luke's use of Mark. We saw in the passages examined in Chapter 7 that proponents of the two-source hypothesis could, in each of the many cases of apparent dependence between the texts of Matthew and Luke in their use of Mark, by some means or other, often by an appeal to the influence of a common oral tradition or common editorial criteria, sometimes by finding evidence of textual corruption, find a way of explaining how this has arisen, without assuming that Luke had Matthew as a source, or vice versa. But in the end, we are still left with a challenge to supporters of the two-source hypothesis. Are they able to deal in a convincing way with the cumulative force of the evidence that appears again and again of seemingly strong and widespread statistical dependence between the texts of Matthew and Luke in their use of Mark? If we are willing to accept that either Luke or Matthew had the other as a source, such questions fall away. Instead, the question arises as to whether there is any need to postulate a Q source to account for the double tradition sections of text that are present in Matthew and Luke but absent from Mark. This is not a question that we examine in any depth in this monograph, but we may simply note again that the analysis in Chapter 2 of the triple-link models showed that it was possible to find models without Q that appeared to fit our data reasonably well.

In view of the arguments that can be mustered by defenders of the two-source hypothesis, it would perhaps be overstating the case to claim that our data on verbal agreements and our statistical analysis provides overwhelming evidence against the validity of the two-source hypothesis and calls into question the existence of a written Q source, but it is the case, to put it more modestly, that our results sit very uncomfortably with the two-source hypothesis, specifically with that aspect of it that assumes that Matthew and Luke worked independently in their use of Mark. Given the notorious intractability of the synoptic problem and the number of different models that are still being advocated, none of them without its deficiencies in explaining the relationships between the synoptic gospels, it should not be surprising that we are unable to come up with any more definite conclusions.

8.2 Directions for future work

If we reject the two-source hypothesis and the existence of a written Q source, we may wish to explore further the alternative hypotheses embodied in the two cases of the triple-link model that assume Markan priority: (i) that Matthew used Mark, but Luke used both Mark and Matthew and (ii) that Luke used Mark, but Matthew used both Mark and Luke. We may first note the results of Chapter 2 that the first of these alternatives, which corresponds to the Farrer hypothesis, gives a somewhat better fit to the data. However, the analyses of Chapters 3 and 5 were inconclusive as to whether it was more likely that Luke had Matthew as a source or Matthew had Luke as a source.

A deeper exploration of these two possibilities would require the construction of a more complex database where the structure of the triple-link model would somehow have to be combined with a binary time series approach. In trying to grapple with how a third author is reading and using the material of the previous two, there will be many difficulties. For example, some will be caused by the fact that the material shared by two or three of the gospels may vary in the order in which it appears in each gospel, sometimes at the level of whole pericopes that are in different positions, sometimes at the level of rearrangement of word order within pericopes, and sometimes by similar pieces of text appearing in different pericopes. So the position of a word in the text often moves from one gospel to another. A simple first step, but nevertheless time-consuming, would be the construction of a binary time series similar in form to the ones used in the present book but using Luke as the base text instead of Mark, as in Chapters 3 and 5, or Matthew, as in Chapter 6.

Naturally, although the assumption of Matthean priority has relatively few advocates nowadays, a similar approach could be used to explore the Augustinian and Griesbach hypotheses. The hypotheses according to which Luke came first, though scarcely ever advocated, could be examined too.

At a basic level, there is scope for validating the data sets constructed for this study, checking for errors in Farmer's colour coding or in the present author's transcription into the binary series. In view of the discussion of Section 6.2, this would be especially beneficial for the data set for Mark. It might be worthwhile recording variant possibilities in cases where the decision over whether there is a verbal agreement or not is a difficult one. Furthermore, variant data sets could be constructed using other versions of the Greek text.

However, as remarked in Section 1.4.2, minor variations in the data sets are unlikely to affect significantly the results of the analysis.

An issue that often arises in discussions about the laborious process of constructing the time series data sets of verbal agreements is whether there are computer programs available for textual analysis that would make the process automatic. For a variety of reasons, and especially because of differences in order from gospel to gospel in which sections of text and individual words are arranged and because of the existence of doublets, the specification of verbal agreements is not always easy and can involve more or less subjective judgments. The translation of this process into a computer program is clearly not a trivial task, but this may be a useful and interesting avenue to explore.

Following in the footsteps of earlier statistical work on the synoptic problem, the present study has been based upon the recording word by word of verbal agreements, using a strict definition of verbal agreement, as discussed in Section 1.4.2. This is a good starting point, as the data sets can be constructed in a fairly straightforward way, even if this is a laborious process and there are some marginal cases where a difficult decision has to be made as to whether a verbal agreement should be recorded or not. But this is clearly a very simplistic approach to the problem. There is enormous scope for exploring ways of developing more sophisticated databases of the texts of the synoptic gospels, their grammatical and narrative structures and their inter-relationships and then developing statistical tools for their analysis, going far beyond the simple idea of just recording which words are retained unchanged from one gospel to another. Such an enterprise would, however, require major inter-disciplinary collaboration and substantial resources of time and manpower.

Finally, it is quite conceivable that the methods developed here might be applicable to other situations where there are a number of documents, similar yet different, whose relationships are to be explored. So there is potential for applications beyond the sphere of the synoptic problem and biblical literature.

A

R code for hidden Markov models

A.1 Maximum likelihood estimation

The material in this section adapts the R code in Appendix A.1 of Zucchini and MacDonald (2009) to the maximum likelihood fitting of an HMM for binary data, including logistic regression on values of covariates and reserving State 1 of the hidden Markov chain to be a state in which the values of the observed process $\{X_t\}$ are necessarily 0, so that $\pi_t^1 = 0$.[1]

In the function logistic.HMM0.pn2pw, Equation (4.30) is used to convert the natural parameters to a vector of working parameters.

```
logistic.HMM0.pn2pw <- function(m,k,gamma,beta)
# Transforms natural to working parameters for a binary-HMM
# with logistic regression
# and State 1 associated with probability 0;
# m hidden states (m > 1), k covariates;
# gamma an m x m transition matrix;
# beta a (k+1) x (m-1) matrix of
# regression coefficients for the other states.
{
    tau <- log(gamma/diag(gamma))
    tgamma <- tau[!diag(m)]
    # vector of non-diagonal elements of tau
    tbeta <- as.vector(beta)
    parvect <- c(tgamma,tbeta)
    parvect
}
```

In the function logistic.HMM0.pw2pn, Equation (4.33) is used to convert a vector of working parameters to the natural parameters. Additionally, the stationary distribution $\boldsymbol{\delta}$ that corresponds to the transition matrix $\boldsymbol{\Gamma}$ is calculated, using the result of Theorem 3 in Section 4.4.3.

```
logistic.HMM0.pw2pn <- function(m,k,parvect)
# Transforms working to natural parameters for a binary-HMM
# with logistic regression
```

[1]Listings of all the code in this appendix, together with the binary time series used in this monograph, may also be found at www.ems.bbk.ac.uk/faculty/abakuks/synoptic.

```
# and State 1 associated with probability 0;
# m hidden states (m > 1), k covariates;
# parvect the vector of working parameters;
# Calculates the stationary distribution for the Markov chain.
{
    gamma <- diag(m)
    gamma[!gamma] <- exp(parvect)[1:(m*(m-1))]
    # fills in non-diagonal elements of gamma
    gamma <- gamma/apply(gamma,1,sum)
    beta <- matrix(parvect[(m*(m-1)+1):((m-1)*(m+k+1))],nrow=k+1)
    delta <- solve(t(diag(m) - gamma + 1),rep(1,m))
    list(gamma=gamma,beta = beta,delta=delta)
}
```

In the function `logistic.HMM0.mllk`, the algorithm described in Section 4.4.1 is used to compute minus the log-likelihood.

```
logistic.HMM0.mllk <- function(m,k,parvect,x,z)
# Computes minus the log-likelihood for a stationary binary-HMM
# with logistic regression
# and State 1 associated with probability 0;
# m hidden states (m > 1), k covariates;
# parvect a vector of working parameters;
# x an n x 1 vector of binary observations;
# z an n x (k+1) design matrix for the covariates.
{
    pn <- logistic.HMM0.pw2pn(m,k,parvect)
    n <- length(x)
    theta <- exp(z%*%pn$beta)
    p <- cbind(rep(0,n),theta/(1 + theta))
    dbinary <- function(x,p){return(p*x + (1-p)*(1-x))}
    allprobs <- dbinary(x,p)
    # allprobs an n x m matrix
    lnw <- 0
    phi <- pn$delta
    for (i in 1:n)
    {
        phi <- phi%*%pn$gamma*allprobs[i,]
        sumphi <- sum(phi)
        lnw <- lnw + log(sumphi)
        phi <- phi/sumphi
    }
    mllk <- -lnw
    mllk
}
```

In the function `logistic.HMM0.mle`, the maximum likelihood estimates of the model parameters are computed, using the `nlm` and `logistic.HMM0.mllk` functions, and listed together with other information about the corresponding fitted model, including the criteria described in Section 4.5 for choosing between models. The design matrix z that has to be supplied is the matrix whose first column is a column of 1s and the other columns are the values of the covariate series, both exogenous and endogenous.

```
logistic.HMM0.mle <- function(m,k,gamma0,beta0,x,z)
# ML estimation for a stationary binary-HMM
# with logistic regression
# and State 1 associated with probability 0;
# m hidden states (m > 1), k covariates;
# gamma0 initial values for the m x m transition matrix;
# beta0 initial values for the (k+1) x (m-1) matrix of
# regression coefficients for the other states;
# x an n x 1 vector of binary observations;
# z an n x (k+1) design matrix for the covariates.
{
    parvect0 <- logistic.HMM0.pn2pw(m,k,gamma0,beta0)
    mod <- nlm(logistic.HMM0.mllk,parvect0,m=m,k=k,x=x,z=z)
    # using the nlm function to minimize minus the log-likelihood
    pn <- logistic.HMM0.pw2pn(m,k,mod$estimate)
    # the MLE of the natural parameters
    gamma <- pn$gamma
    delta <- pn$delta
    meanstay <- 1/(delta%*%diag(1 - gamma))
    meanstay <- as.vector(meanstay)
    # expected length of stay in a hidden state
    mllk <- mod$minimum   # the minimum attained
    np <- length(parvect0)
    AIC <- 2*(mllk+np)   # Akaike information criterion
    n <- length(x)
    BIC <- 2*mllk+np*log(n)   # Bayesian information criterion
    if(k==0) {zp <- matrix(c(1),nrow=1)}
    if(k==1) {zp <- matrix(c(1,1,0,1),nrow=2)}
    if(k==2) {zp <- matrix(c(1,1,1,1,0,0,1,1,0,1,0,1),nrow=4)}
    if(k==3) {zp <- matrix(c(1,1,1,1,1,1,1,1,
    0,0,0,0,1,1,1,1,0,0,1,1,0,0,1,1,0,1,0,1,0,1,0,1),nrow=8)}
    if(k<=3)
    {
        theta <- exp(zp%*%pn$beta)
        p <- cbind(rep(0,2^k),theta/(1 + theta))
    }
    if(k>=4) {zp <- p <-  "-"}
    list(gamma=gamma,beta=pn$beta,delta=delta,meanstay=meanstay,
    code=mod$code,mllk=mllk,AIC=AIC,BIC=BIC,zp=zp,p=p)
    # 'code' indicates how nlm terminated;
    # zp is a 2^k x (k+1) matrix, each row a set of possible values
    # of the rows of the design matrix for binary covariates;
    # p is a 2^k x m matrix, with fitted probabilities for each
    # possible row of the design matrix and each hidden state.
}
```

A.2 Decoding

The material in this section adapts some of the R code in Appendix A.2 of Zucchini and MacDonald (2009) to decode an observed binary series.

Given a fitted HMM, local decoding for each time point t finds the most likely hidden state i_t to have given rise to the observed series. The functions `logistic.HMM0.lalphabeta`, `logistic.HMM0.stateprobs` and `logistic.HMM0.localdecoding` implement the method of local decoding described in Section 4.6.1.

The logarithms of the forward probabilities $\boldsymbol{\alpha}_t$ and the backward probabilities $\boldsymbol{\beta}_t$ are calculated in the function `logistic.HMM0.lalphabeta` using the scaling procedures described in Sections 4.4.1 and 4.6.2. To avoid any possible confusion over notation, it should be observed that `beta` in the R code is used for the matrix of regression coefficients β_{ji} as they appear in Equation (4.10). In the functions used for local decoding, the matrices `lbeta` and `lb` are used to store the logarithms of the backwards probabilities $\boldsymbol{\beta}_t$.

```
logistic.HMM0.lalphabeta<-function(m,k,gamma,beta,x,z)
# Computes the logarithms of the forward and backward
# probabilities in the form of m x n matrices
# for a given stationary binary-HMM
# with logistic regression
# and State 1 associated with probability 0;
# m hidden states (m > 1), k covariates;
# gamma an m x m transition matrix;
# beta a (k+1) x (m-1) matrix of
# regression coefficients for the other states;
# x an n x 1 vector of binary observations;
# z an n x (k+1) design matrix for the covariates.
{
    delta <- solve(t(diag(m)-gamma+1),rep(1,m))
    n <- length(x)
    lalpha <- lbeta <- matrix(NA,m,n)
    theta <- exp(z%*%beta)
    p <- cbind(rep(0,n),theta/(1 + theta))
    dbinary <- function(x,p){return(p*x + (1-p)*(1-x))}
    allprobs <- dbinary(x,p)
    # allprobs an n x m matrix
    lnw <- 0
    phi <- delta
    for (i in 1:n)
    {
        phi <- phi%*%gamma*allprobs[i,]
        sumphi <- sum(phi)
        lnw <- lnw + log(sumphi)
        phi <- phi/sumphi
```

```
        lalpha[,i] <- log(phi) + lnw
    }
    lbeta[,n]  <- rep(0,m)
    psi <- rep(1/m,m)
    lns <- log(m)
    for (i in (n-1):1)
    {
        psi <- gamma%*%(allprobs[i+1,]*psi)
        lbeta[,i] <- log(psi) + lns
        sumpsi <- sum(psi)
        psi <- psi/sumpsi
        lns <- lns + log(sumpsi)
    }
    list(la=lalpha,lb=lbeta)
}
```

Given the observed series and a fitted HMM, the conditional probability distribution for the hidden state at each time point t is found by the function `logistic.HMM0.stateprobs`.

```
logistic.HMM0.stateprobs <- function(m,k,gamma,beta,x,z)
# Calculates the conditional state probabilities
# in the form of an m x n matrix
# for a given stationary binary-HMM
# with logistic regression
# and State 1 associated with probability 0;
# m hidden states (m > 1), k covariates;
# gamma an m x m transition matrix;
# beta a (k+1) x (m-1) matrix of
# regression coefficients for the other states;
# x an n x 1 vector of binary observations;
# z an n x (k+1) design matrix for the covariates.
{
    delta <- solve(t(diag(m)-gamma+1),rep(1,m))
    n <- length(x)
    fb <- logistic.HMM0.lalphabeta(m,k,gamma,beta,x,z)
    la <- fb$la
    lb <- fb$lb
    c <- max(la[,n])
    # c introduced to reduce the chances of underflow
    llk <- c+log(sum(exp(la[,n]-c)))
    stateprobs <- matrix(NA,ncol=n,nrow=m)
    for (i in 1:n) stateprobs[,i] <- exp(la[,i]+lb[,i]-llk)
    stateprobs
}
```

Given the observed series and a fitted HMM, the most likely hidden state i_t at each time point t is found by the function `logistic.HMM0.localdecoding`, using the output from the function `logistic.HMM0.stateprobs`.

```
logistic.HMM0.localdecoding <- function(m,k,gamma,beta,x,z)
# Performs local decoding
# for a given stationary binary-HMM
# with logistic regression
# and State 1 associated with probability 0;
# m hidden states (m > 1), k covariates;
# gamma an m x m transition matrix;
# beta a (k+1) x (m-1) matrix of
# regression coefficients for the other states;
# x an n x 1 vector of binary observations;
# z an n x (k+1) design matrix for the covariates.
{
    n <- length(x)
    stateprobs <- logistic.HMM0.stateprobs(m,k,gamma,beta,x,z)
    ild <- rep(NA,n)
    for (i in 1:n) ild[i]<-which.max(stateprobs[,i])
    ild
}
```

Given a fitted HMM, global decoding finds the most likely sequence of hidden states to have given rise to the observed series. The function `logistic.HMM0.viterbi` implements the method of global decoding using the Viterbi algorithm as described in Section 4.6.3.

```
logistic.HMM0.viterbi<-function(m,k,gamma,beta,x,z)
# Performs global decoding using the Viterbi algorithm
# for a given stationary binary-HMM
# with logistic regression
# and State 1 associated with probability 0;
# m hidden states (m > 1), k covariates;
# gamma an m x m transition matrix;
# beta a (k+1) x (m-1) matrix of
# regression coefficients for the other states;
# x an n x 1 vector of binary observations;
# z an n x (k+1) design matrix for the covariates.
{
    delta <- solve(t(diag(m)-gamma+1),rep(1,m))
    n <- length(x)
    theta <- exp(z%*%beta)
    p <- cbind(rep(0,n),theta/(1 + theta))
    dbinary <- function(x,p){return(p*x + (1-p)*(1-x))}
    allprobs <- dbinary(x,p)
    # allprobs an n x m matrix
    xi <- matrix(0,n,m)
    phi <- delta*allprobs[1,]
    xi[1,] <- phi/sum(phi)
    # xi scaled to have row sums one
    # in order to avoid problems of underflow.
    for (i in 2:n)
```

```
    {
        phi <- apply(xi[i-1,]*gamma,2,max)*allprobs[i,]
        xi[i,] <- phi/sum(phi)
    }
    iv <- numeric(n)
    iv[n] <- which.max(xi[n,])
    for (i in (n-1):1)
    iv[i] <- which.max(gamma[,iv[i+1]]*xi[i,])
    iv
}
```

Bibliography

Abakuks, A. (2006a). A statistical study of the triple-link model in the synoptic problem. *J. R. Statist. Soc. A 169*, 49–60.

Abakuks, A. (2006b). The synoptic problem and statistics. *Significance 3*, 153–157.

Abakuks, A. (2007). A modification of Honoré's triple-link model in the synoptic problem. *J. R. Statist. Soc. A 170*, 841–850.

Abakuks, A. (2012). The synoptic problem: on Matthew's and Luke's use of Mark. *J. R. Statist. Soc. A 175*, 959–975.

Abakuks, A. (in press, 2015). A statistical time series approach to the use of Mark by Matthew and Luke. In J. C. Poirier and J. Peterson (Eds.), *Marcan Priority without Q: Explorations in the Farrer Hypothesis.* London: Bloomsbury, T & T Clark.

Adams, E. (2011). *Parallel Lives of Jesus: A Narrative-Critical Guide to the Four Gospels.* London: SPCK.

Aland, K. (Ed.) (1971). *Synopsis Quattuor Evangeliorum* (7th ed.). Stuttgart: Würtembergische Bibelanstalt.

Aland, K. (Ed.) (1996). *Synopsis Quattuor Evangeliorum* (15th ed.). Stuttgart: Deutsche Bibelgesellschaft.

Barr, A. (1995). *A Diagram of Synoptic Relationships* (2nd ed.). Edinburgh: T & T Clark.

Bartholomew, D. J. (1988). Probability, statistics and theology. *J. R. Statist. Soc. A 151*, 137–178.

Bartholomew, D. J. (1996). *Uncertain Belief: Is It Rational to Be a Christian?* Oxford: Clarendon Press.

Bartholomew, D. J., M. Knott, and I. Moustaki (2011). *Latent Variable Models and Factor Analysis: A Unified Approach* (3rd ed.). Chichester: Wiley.

Bauckham, R. (2006). *Jesus and the Eyewitnesses: The Gospels as Eyewitness Testimony.* Grand Rapids: Eerdmans.

Bellinzoni, A. J. (Ed.) (1985). *The Two-Source Hypothesis: A Critical Appraisal.* Macon: Mercer University Press.

Burkett, D. R. (2004). *Rethinking the Gospel Sources: From Proto-Mark to Mark.* London: T & T Clark.

Burkett, D. R. (2009). *The Unity and Plurality of Q*, Volume 2 of *Rethinking the Gospel Sources.* Atlanta: Society of Biblical Literature.

Burridge, R. A. (2004). *What Are the Gospels?: A Comparison with Graeco-Roman Biography* (2nd ed.). Grand Rapids: Eerdmans.

Burridge, R. A. (2005). *Four Gospels, One Jesus?* (2nd ed.). London: SPCK.

Butler, B. C. (1939). St. Luke's debt to St. Matthew. *Harv. Theol. Rev. 32*, 237–308.

Butler, B. C. (1951). *The Originality of St. Matthew: A Critique of the Two-Document Hypothesis.* Cambridge: Cambridge University Press.

Carlston, C. E. and D. Norlin (1971). Once more — statistics and Q. *Harv. Theol. Rev. 64*, 59–78.

Carlston, C. E. and D. Norlin (1999). Statistics and Q — some further observations. *Nov. Test. 41*, 108–123.

Catchpole, D. R. (1993). *The Quest for Q.* Edinburgh: T & T Clark.

Crook, Z. A. (2012). *Parallel Gospels: A Synopsis of Early Christian Writing.* New York: Oxford University Press.

Davies, W. D. and D. C. Allison (1988-1997). *A Critical and Exegetical Commentary on the Gospel according to Saint Matthew.* Edinburgh: T & T Clark. (3 Volumes).

Dines, J. M. (2004). *The Septuagint.* London: Continuum.

Downing, F. G. (1988). Compositional conventions and the synoptic problem. *J. Bibl. Lit. 107*, 69–85.

Downing, F. G. (1992). A paradigm perplex: Luke, Matthew and Mark. *New Test. Stud. 38*, 15–36.

Downing, F. G. (2000). *Doing Things with Words in the First Christian Century.* Sheffield: Sheffield Academic Press.

Dungan, D. L. (1999). *A History of the Synoptic Problem: The Canon, the Text, the Composition, and the Interpretation of the Gospels.* New York: Doubleday.

Dunn, J. D. G. (2003a). Altering the default setting: re-envisaging the early transmission of the Jesus tradition. *New Test. Stud. 49*, 139–175.

Dunn, J. D. G. (2003b). *Jesus Remembered*, Volume 1 of *Christianity in the Making*. Grand Rapids: Eerdmans.

Dunn, J. D. G. (2013). *The Oral Gospel Tradition*. Grand Rapids: Eerdmans.

Ehrman, B. D. and Z. Pleše (2011). *The Apocryphal Gospels: Texts and Translations*. New York: Oxford University Press.

Eusebius (1926). *Ecclesiastical History: Books I-V, translated by Kirsopp Lake*. Loeb Classical Library. Cambridge, Massachusetts: Harvard University Press.

Eve, E. (2013). *Behind the Gospels: Understanding the Oral Tradition*. London: SPCK.

Farmer, W. R. (1964). *The Synoptic Problem: a Critical Analysis*. London: Collier-Macmillan.

Farmer, W. R. (1969). *Synopticon: The Verbal Agreement between the Greek Texts of Matthew, Mark and Luke Contextually Exhibited*. Cambridge: Cambridge University Press.

Farmer, W. R. (1994). *The Gospel of Jesus: The Pastoral Relevance of the Synoptic Problem*. Louisville: Westminster/John Knox Press.

Farrer, A. M. (1955). On dispensing with Q. In D. E. Nineham (Ed.), *Studies in the Gospels: Essays in Memory of R. H. Lightfoot*, pp. 55–88. Oxford: Blackwell.

Foster, P. (Ed.) (2008). *The Non-Canonical Gospels*. London: T & T Clark.

France, R. T. (2002). *The Gospel of Mark: A Commentary on the Greek Text*. Carlisle: Paternoster Press.

Gathercole, S. (2012). *The Composition of the Gospel of Thomas: Original Language and Influences*. Cambridge: Cambridge University Press.

Goodacre, M. (1996). *Goulder and the Gospels: An Examination of a New Paradigm*. Sheffield: Sheffield Academic Press.

Goodacre, M. (1999). Beyond the Q impasse or down a blind alley? *J. Study New Test. 76*, 33–52.

Goodacre, M. (2001). *The Synoptic Problem: A Way through the Maze*. London: Sheffield Academic Press.

Goodacre, M. (2002). *The Case against Q: Studies in Markan Priority and the Synoptic Problem*. Harrisburg: Trinity Press International.

Goodacre, M. (2012). *Thomas and the Gospels: The Making of an Apocryphal Text*. London: SPCK.

Goodacre, M. and N. Perrin (Eds.) (2004). *Questioning Q*. London: SPCK.

Goulder, M. D. (1974). *Midrash and Lection in Matthew*. London: SPCK.

Goulder, M. D. (1978). On putting Q to the test. *New Test. Stud. 24*, 218–234.

Goulder, M. D. (1989). *Luke: A New Paradigm*. Sheffield: Sheffield Academic Press.

Granger, C. W. J. (1969). Investigating causal relations by econometric models and cross-spectral methods. *Econometrica 37*, 424–438.

Grant, R. M. (1997). *Irenaeus of Lyons*. London: Routledge.

Greeven, H. (1978). The gospel synopsis from 1776 to the present day. In B. Orchard and T. R. W. Longstaff (Eds.), *J. J. Griesbach: Synoptic and Text-Critical Studies 1776-1976*, pp. 22–49. Cambridge: Cambridge University Press.

Grimmett, G. R. and D. R. Stirzaker (2001). *Probability and Random Processes* (3rd ed.). Oxford: Oxford University Press.

Gundry, R. H. (1994). *Matthew: A Commentary on His Handbook for a Mixed Church under Persecution* (2nd ed.). Grand Rapids: Eerdmans.

Hartzel, J., A. Agresti, and B. Caffo (2001). Multinomial logit random effects models. *Statist. Modllng. 1*, 81–102.

Hawkins, J. C. (1909). *Horae Synopticae: Contributions to the Study of the Synoptic Problem* (2nd ed.). Oxford: Clarendon Press.

Hengel, M. (2000). *The Four Gospels and the One Gospel of Jesus Christ*. London: SCM Press.

Holmes, D. I. (1998). The evolution of stylometry in humanities scholarship. *Literary and Linguistic Computing 13*, 111–117.

Honoré, A. M. (1968). A statistical study of the synoptic problem. *Nov. Test. 10*, 95–147.

Hooker, M. D. (1991). *The Gospel According to St Mark*. London: A & C Black.

Huck, A. (1949). *Synopsis of the First Three Gospels* (9th ed.). Oxford: Blackwell.

Huggins, R. V. (1992). Matthean posteriority: a preliminary proposal. *Nov. Test. 34*, 1–22.

Humphreys, C. J. (2011). *The Mystery of the Last Supper: Reconstructing the Final Days of Jesus*. Cambridge: Cambridge University Press.

Hurtado, L. W. (2006). *The Earliest Christian Artifacts: Manuscripts and Christian Origins.* Grand Rapids: Eerdmans.

Kedem, B. and K. Fokianos (2002). *Regression Models for Time Series Analysis.* Hoboken: Wiley.

Kenny, A. (1986). *A Stylometric Study of the New Testament.* Oxford: Clarendon Press.

Kloppenborg, J. S. (1987). *The Formation of Q: Trajectories in Ancient Wisdom Collections.* Philadelphia: Fortress Press.

Kloppenborg, J. S. (2000). *Excavating Q: The History and Setting of the Sayings Gospel.* Minneapolis: Fortress Press.

Kloppenborg, J. S. (2008). *Q, the Earliest Gospel: An Introduction to the Original Stories and Sayings of Jesus.* Louisville: Westminster John Knox Press.

Law, T. M. (2013). *When God Spoke Greek: The Septuagint and the Making of the Christian Bible.* New York: Oxford University Press.

Ledger, G. R. (1995). An exploration of differences in the Pauline epistles using multivariate statistical analysis. *Lit. Ling. Comput. 10*, 85–97.

Mächler, M. and P. Bühlmann (2004). Variable length Markov chains: methodology, computing, and software. *J. of Computnl and Graph. Statist. 13*, 435–455.

Marshall, I. H. (1978). *The Gospel of Luke: A Commentary on the Greek Text.* Exeter: Paternoster Press.

Matilla, S. L. (1994). A problem still clouded: yet again — statistics and "Q". *Nov. Test. 36*, 313–329.

Matilla, S. L. (2004). Negotiating the clouds around statistics and "Q": A rejoinder and independent analysis. *Nov. Test. 46*, 105–131.

McCullagh, P. and J. A. Nelder (1989). *Generalized Linear Models* (2nd ed.). London: Chapman & Hall.

McNicol, A. J., D. L. Dungan, and D. B. Peabody (Eds.) (1996). *Beyond the Q Impasse — Luke's Use of Matthew.* Valley Forge: Trinity Press International.

Mealand, D. L. (1996). The extent of the Pauline corpus: A multivariate approach. *J. Study New Test. 18*, 61–92.

Mealand, D. L. (2011). Is there stylometric evidence for Q? *New Test. Stud. 57*, 483–507.

Metzger, B. M. and B. D. Ehrman (2005). *The Text of the New Testament: Its Transmission, Corruption, and Restoration* (4th ed.). New York: Oxford University Press.

Michaelson, S. and A. Q. Morton (1972). Last words: A test of authorship for Greek writers. *New Test. Stud. 18*, 178–191.

Morgenthaler, R. (1971). *Statistische Synopse.* Zürich/Stuttgart: Gotthelf-Verlag.

Morton, A. Q. (1965). The authorship of Greek prose. *J. R. Statist. Soc. A 128*, 169–233.

Morton, A. Q. (1978). *Literary Detection: How to Prove Authorship and Fraud in Literature and Documents.* Epping: Bowker.

Mosconi, R. and R. Seri (2006). Non-causality in bivariate binary time series. *J. Econometr. 132*, 379–407.

Mosteller, F. and D. L. Wallace (1984). *Applied Bayesian and Classical Inference: The Case of The Federalist Papers* (2nd ed.). New York: Springer-Verlag.

Neirynck, F. (Ed.) (1974). *The Minor Agreements of Matthew and Luke against Mark with a Cumulative List.* Leuven: Leuven University Press.

Neirynck, F. (1991). *The Minor Agreements in a Horizontal-Line Synopsis.* Leuven: Peeters.

Nestle-Aland (Ed.) (1963). *Novum Testamentum Graece* (25th ed.). Stuttgart: Würtembergische Bibelanstalt.

Nestle-Aland (Ed.) (2012). *Novum Testamentum Graece* (28th ed.). Stuttgart: Deutsche Bibelgesellschaft.

Neumann, K. J. (1990). *The Authenticity of the Pauline Epistles in the Light of Stylostatistical Analysis.* Atlanta: Scholars Press.

Neville, D. J. (2002). *Mark's Gospel – Prior or Posterior?: A Reappraisal of the Phenomenon of Order.* Sheffield: Sheffield Academic Press.

Oakes, M. P. (2014). *Literary Detective Work on the Computer.* Amsterdam: John Benjamins.

Orchard, B. and T. R. W. Longstaff (Eds.) (1978). *J. J. Griesbach: Synoptic and Text-Critical Studies 1776-1976.* Cambridge: Cambridge University Press.

O'Rourke, J. J. (1974). Some observations on the synoptic problem and the use of statistical procedures. *Nov. Test. 16*, 272–277.

Palmer, N. H. (1967). Lachmann's argument. *New Test. Stud. 13*, 368–378.

Parker, D. C. (2010). *Codex Sinaiticus: The Story of the World's Oldest Bible.* London: British Library.

Peabody, D. B. (1983). Augustine and the Augustinian hypothesis: A reexamination of Augustine's thought in *De consensu evangelistarum*. In W. R. Farmer (Ed.), *New Synoptic Studies: The Cambridge Gospel Conference and Beyond*, pp. 37–64. Macon: Mercer University Press.

Peabody, D. B., L. Cope, and A. J. McNicol (Eds.) (2002). *One Gospel from Two: Mark's Use of Matthew and Luke.* Harrisburg: Trinity Press International.

Pietersma, A. and B. G. Wright (Eds.) (2007). *A New English Translation of the Septuagint.* New York: Oxford University Press.

Poirier, J. C. (2008). Statistical studies of the verbal agreements and their impact on the synoptic problem. *Currs. Bibl. Res. 7*, 68–123.

Poirier, J. C. (2012). The roll, the codex, the wax tablet and the synoptic problem. *J. Study New Test. 35*, 3–30.

Poirier, J. C. and J. Peterson (Eds.) (in press, 2015). *Marcan Priority without Q: Explorations in the Farrer Hypothesis.* London: Bloomsbury, T & T Clark.

Rahlfs, A. (Ed.) (1979). *Septuaginta.* Stuttgart: Deutsche Bibelgesellschaft.

Robinson, J. M., P. Hoffmann, and J. S. Kloppenborg (Eds.) (2000). *The Critical Edition of Q.* Minneapolis: Fortress Press.

Sanday, W. (1911). The conditions under which the gospels were written, in their bearing upon some difficulties of the synoptic problem. In W. Sanday (Ed.), *Studies in the Synoptic Problem*, pp. 3–26. Oxford: Clarendon Press.

Smith, D. M. (2001). *John among the Gospels* (2nd ed.). Columbia: University of South Carolina Press.

Stamatatos, E. (2009). A survey of modern authorship attribution methods. *J. Am. Soc. Inform. Sci. Tec. 60*, 538–556.

Stanton, G. (2002). *The Gospels and Jesus* (2nd ed.). New York: Oxford University Press.

Stein, R. H. (2001). *Studying the Synoptic Gospels: Origin and Interpretation* (2nd ed.). Grand Rapids: Baker Academic.

Stram, D. O. and J. W. Lee (1994). Variance components testing in the longitudinal mixed effects model. *Biometrics 50*, 1171–1177.

Streeter, B. H. (1924). *The Four Gospels: A Study of Origins.* London: Macmillan.

Throckmorton, B. H. (Ed.) (1992). *Gospel Parallels: A Comparison of the Synoptic Gospels* (5th ed.). Nashville: Thomas Nelson Publishers.

Tuckett, C. M. (1983). *The Revival of the Griesbach Hypothesis: An Analysis and Appraisal.* Cambridge: Cambridge University Press.

Tuckett, C. M. (1984). On the relationship between Matthew and Luke. *New Test. Stud. 30*, 130–142.

Tuckett, C. M. (1987). *Reading the New Testament: Methods of Interpretation.* London: SPCK.

Tuckett, C. M. (1996). *Q and the History of Early Christianity: Studies on Q.* Edinburgh: T & T Clark.

Tyson, J. B. and T. R. W. Longstaff (1978). *Synoptic Abstract*, Volume XV of *The Computer Bible.* Wooster: Biblical Research Associates.

Vinson, R. (2004). How minor? Assessing the significance of the minor agreements as an argument against the two-source hypothesis. In M. Goodacre and N. Perrin (Eds.), *Questioning Q*, pp. 151–164. London: SPCK.

Vinzent, M. (2011). *Christ's Resurrection in Early Christianity and the Making of the New Testament.* Farnham: Ashgate.

Vinzent, M. (2014). *Marcion and the Dating of the Synoptic Gospels.* Studia Patristica Supplement 2. Leuven: Peeters.

Visscher, P. M. (2006). A note on the asymptotic distribution of likelihood ratio tests to test variance components. *Twin Res. Hum. Genet. 9*, 490–495.

Watson, F. (2009). Q as hypothesis: A study in methodology. *New Test. Stud. 55*, 397–415.

Watson, F. (2013). *Gospel Writing: A Canonical Perspective.* Grand Rapids: Eerdmans.

Wenham, D. (1972). The synoptic problem revisited: some new suggestions about the composition of Mark 4:1-34. *Tynd. Bull. 23*, 3–38.

Wenham, J. W. (1991). *Redating Matthew, Mark and Luke: A Fresh Assault on the Synoptic Problem.* Sevenoaks: Hodder & Stoughton.

Wevers, J. W. (Ed.) (1977). *Deuteronomium*, Volume III,2 of *Septuaginta:Vetus Testamentum Graecum.* Göttingen: Vandenhoeck & Ruprecht.

Yee, T. W. and C. J. Wild (1996). Vector generalized additive models. *J. R. Statist. Soc. B 58*, 481–493.

Zucchini, W. and I. L. MacDonald (2009). *Hidden Markov Models for Time Series: An Introduction Using R*. Boca Raton: Chapman & Hall/CRC Press.

Author Index

Subject Index